Advance Praise for

THE END

· · · · ·

"The 21st century may well be the most important epoch in our species' history, in which we either survive and flourish in centuries to come, or trigger our own extinction. Phil Torres' *The End* is **the most chilling reality check on over-optimism I've ever read**, a gripping narrative that could be a Hollywood blockbuster film script but in fact is grounded in facts and data about what threatens to send humanity the way of the dinosaurs. Expect the best but prepare for the worst by reading this important book."

—**Michael Shermer**, Founding Publisher of *Skeptic* magazine, monthly columnist for *Scientific American*, and author of *The Moral Arc: How Science and Reason Lead Humanity toward Truth, Justice, and Freedom*

"In atheism, as in many religious traditions, the end of the world has a special significance. For us, the goal is to make sure it doesn't happen. **Phil Torres' book is a great start**. It's a provocative look at existential risks near and far, and is sure to get people thinking about these important questions."

—**Sean Carroll**, theoretical physicist at Caltech and author of *The Particle at the End of the Universe*

"Secular Westerners who study catastrophic risks largely ignore the fact that, for the majority of people on the planet, the risks to the future of human existence are seen through a religious lens. **Torres makes an enormous contribution** with this explanation of the relationship between those eschatological belief systems and attempts to create a science of catastrophic risk estimation around threats from both technology and the natural world. . . . People interested in the apocalyptic worldviews of groups like ISIS and the Christian Right need to read this book to understand how contemporary science will be woven into the eschatological narrative. People interested in ensuring that humanity successfully navigates the 21st century need to read this concise and comprehensive contribution to catastrophic risk literature."

—**James Hughes**, Executive Director, Institute for Ethics and Emerging Technologies and author of *Citizen Cyborg*

"**Thoughtful, incisive, scholarly**—and a pleasure to read."

—**David Pearce**, Oxford philosopher, co-founder of Humanity+, and author of *The Hedonistic Imperative*

"If science and religion agree on nothing else, they agree on one singularly important thing: the world will one day come to an end. Any agreement stops there. **As Phil Torres compellingly and forcefully argues**, the surest way to avoid—or rather, delay—this inevitability, is to remove religion from the entire conversation. Only then will we begin to have a chance of accurately setting and slowing the Doomsday Clock."

—**Peter Boghossian**, author of *A Manual for Creating Atheists*

"Phil Torres takes us on **a fascinating journey through some rather alarming territory**. His fear that the human race may soon be extinct is all too clearly justified—but so also is his message that we can do many things to minimize the danger."

—**John Leslie**, Fellow of the Royal Society of Canada, and author of *The End of the World: The Science and Ethics of Human Extinction*

"Here's how the world will end—let me count the ways. Wait, no need; **Phil Torres has enumerated—and analyzed—the bulk of them in his masterful new book *The End***. With the realization that human beings are unplanned, undersigned, and solely responsible for ourselves comes another recognition: There's no guarantee how long the human adventure will continue. Torres examines a broad selection of ways it could all come crashing down, be it next week or after generations—**and he makes what could be a grim journey as entertaining as it is enlightening**. If humankind is to avoid any of these possible apocalypses, it will have to engineer its own (pardon the expression) salvation. Of course that is doubly challenged when so many of the possible threats are of our own making! Risk analysis of the caliber Torres demonstrates in *The End* may well equip us to respond more capably to the next existential crisis we face, no matter what it turns out to be."

—**Tom Flynn**, editor of *Free Inquiry*, Executive Director of the Council for Secular Humanism, and editor of *The New Encyclopedia of Unbelief*

"I love this book. **Torres is a rare communicator who handles big topics with extraordinary skill**. In this stunning, scary, and fun book, he tackles nothing less than the end of us. From irrational religious claims about one apocalypse or another to sober, science-based possibilities of real doomsdays, Torres covers it all in grand style. This must-read book answers many questions while also spurring readers to think of new questions—exactly what a great book should do. Don't hesitate. Hurry and read *The End* before the clock strikes twelve!"

—**Guy P. Harrison**, author of *Good Thinking: What You Need to Know to Be Smarter, Safer, Wealthier, and Wiser* and *50 Simple Questions for Every Christian*

"Torres does **an excellent job introducing the potential dangers of superintelligent machines**, without expecting the reader to be an expert in artificial intelligence, but without dumbing down the topic."

—**Roman Yampolskiy**, Director of the Cybersecurity Laboratory at the University of Louisville and 2015 Research Advisor at the Machine Intelligence Research Institute

"Phil Torres' **interesting and highly original book** incorporates religion, science, and thoughtful science fiction (the kind that might someday become science). He illuminates the dangers of a religious apocalypse (laughable, were it not for the possibility that believers in a fictional, god-induced apocalypse might acquire the scientific technology to bring about a nonfictional, human-induced apocalypse). Warning of inherent worldwide dangers in the increasing number of people who dismiss science and don't care to distinguish fantasy from reality, **Torres motivates atheists to go beyond simply criticizing and making fun of fundamentalist religions**. Though generally pessimistic about the future of our planet, Torres provides the reader with strategies to maximize the likelihood that future generations will continue to have a quality life on earth—or, if necessary, on another planet."

—**Herb Silverman**, Distinguished Professor Emeritus of the College of Charleston, Founder and President of the Secular Coalition for America, and author of *Candidate Without a Prayer*

"This one-of-a-kind-book provides an accessible yet expert education into several global doomsday threats, both secular and religious, both real and possible. **Highly enlightening and very highly recommended!**"

—**John W. Loftus**, author of *How to Defend the Christian Faith*

"*The End* is a fascinating read. It provides a detailed account of the biggest issues confronting civilization and their various causes. This is **an important read** for anyone interested in making sure the promises of advanced technologies are fully realized and the perils effectively eliminated."

—**Zoltan Istvan**, author of *The Transhumanist Wager* and 2016 Presidential candidate for the Transhumanist Party

"David Hume famously said that mistakes in religion were dangerous. He didn't know how dangerous, **nor did we until Phil Torres explained it in *The End***."

—**Alex Rosenberg**, R. Taylor Cole Professor of Philosophy at Duke University and author of *The Atheist's Guide to Reality* and *The Girl from Krakow*

THE END

What Science and Religion
Tell Us about the **Apocalypse**

Phil Torres

Foreword by Russell Blackford

PITCHSTONE PUBLISHING
Durham, North Carolina

Pitchstone Publishing
Durham, North Carolina
www.pitchstonepublishing.com

Copyright © 2016 by Phil Torres

Library of Congress Cataloging-in-Publication Data

Names: Torres, Phil, author.
Title: The end : what religion and science tell us about the Apocalypse /
    Phil Torres ; foreword by Russell Blackford.
Description: Durham, North Carolina : Pitchstone Publishing, 2016. | Includes
    bibliographical references.
Identifiers: LCCN 2015027511 | ISBN 9781634310406 (pbk. : alk. paper)
Subjects: LCSH: End of the world. | Eschatology. | Religion and science.
Classification: LCC BL503 .T67 2016 | DDC 001.9—dc23
LC record available at http://lccn.loc.gov/2015027511

*We turned the switch and saw the flashes. We watched them
for a little while and then we switched everything off
and went home. That night, there was very little doubt
in my mind that the world was headed for grief.*
**—Leó Szilárd (after conducting an experiment in 1939
that proved a nuclear chain reaction is possible)**

*I know not with what weapons World War III will be fought,
but World War IV will be fought with sticks and stones.*
—Albert Einstein

Will we ever exhaust the rich supply of dooms?
—David Brin

*Religion will act as an opiate and keep people quiet if you
promise them something in the distant future. When it's
in the immediate future, apocalyptic religiosity is not an opiate,
it's a combination of psychomimetic LSD and crack.*
—Richard Landes

Contents

· · · · ·

Foreword

· · · · ·

About 65 million years ago, the Mesozoic era—the age of dinosaurs—came to a spectacular end. Today, the Cretaceous-Tertiary (K-T) boundary[1] marks the event: it is a thin stratum of dark rock with a strangely high concentration of iridium that appears to derive from an extraterrestrial source. Scientists theorize that an object from space plunged into the earth near modern-day Chicxulub, a town located on the Yucatán Peninsula in southeastern Mexico. The impact must have produced cataclysmic effects: earthquakes, volcanic eruptions, wildfires over vast areas, mountainous tsunamis, and the equivalent of a nuclear winter. It left behind a crater over one hundred miles wide, first detected in the 1960s and 1970s but not announced to the wider scientific community until the early 1980s. Ecological damage on a worldwide scale led to mass extinctions, including the obliteration of nonavian dinosaur species.

The K-T extinction event was the most recent of five that occurred on such a scale. The first, actually a pair of events, took place about 445 million years ago at the end of the Ordovician period. Since then, planetary-level species extinctions have marked the end of the Devonian period (about 360 million years ago), the end of the Permian period, and with it the Paleozoic era (about 250 million years ago), and the end of the Triassic/beginning of the Jurassic (about 200 million years ago). The K-T extinction event is the best known, and in some ways the most dramatic—involving, as it did, the impressive dinosaurs of the late Mesozoic (heyday of *Triceratops* and *Tyrannosaurus rex*, among others). Most importantly, from a human perspective, the extinction of the dinosaurs opened up an evolutionary path that led to *Homo sapiens*.

Arguably, our planet is going through a sixth major extinction event, precipitated, this time, by human conduct, and notable for the loss of

13

many species and huge tracts of natural habitat. Our species is polluting the sky, the land, and the seas, clearing forests, heating the planet, and generally transforming Earth's biosphere. Taken to an extreme—though not an altogether implausible one—we could set off a chain of events that culminates in global ecological collapse, leaving our world uninhabitable.

Ultimate disaster could, however, befall us from external inputs, such as the arrival of an asteroid or a comet set for collision with our precious, pale-blue planet. As happened to the dinosaurs, a hammer could fall on us from space, and with current technology there is nothing we could do to stop it. Alternatively, civilization could be shattered, and our survival as a species threatened, by a volcanic supereruption, such as that of Mount Toba, in Indonesia, about 70,000 years ago.

Throughout *The End: What Science and Religion Tell Us about the Apocalypse*, Phil Torres examines possible causes of humanity's doom. While mainly focused on mechanisms for our total extinction, he also considers desolate futures in which some of us might struggle on in the aftermath of irreversible cultural disintegration. Any of these outcomes would entail the end of *Homo sapiens'* potential for social and scientific advances. Torres offers a comprehensive account of the possibilities: he names and explains the usual suspects in the (possible) fate of humankind. As he also explains, however, the greatest dangers may come from unknown unknowns: they may arise from causes that we cannot, with our current knowledge and intelligence, even imagine.

Advances in technology, and particularly the design and manufacture of enormously destructive weapons, have shaped a world where we could become our own destroyers. The Cold War ended a quarter of a century ago with the fall of the Soviet Union, and there has since been some decommissioning of nuclear warheads. Nonetheless, Russia, the United States of America, and other nuclear powers have stockpiled more than enough hydrogen bombs to smash the fragile bases of modern civilization or ruin the global ecological balance itself. Military researchers continue to devise weapons that draw on new, sophisticated technologies. Advanced nanotech weapons, for example, may become devastatingly effective.

We are entering a geopolitical environment where weapons of mass destruction may fall into the hands of nonstate actors, such as extremist organizations or individual terrorists. To compound the problem, some ideological extremists, terrorists, and even mainstream politicians are motivated by apocalyptic or millennial belief systems. Some fanatics are willing to risk global catastrophe to achieve their goals; as Torres emphasizes, some actively seek to provoke a last-days Battle of Armageddon. The short of it is: our world contains too many apocalyptic ideologues and too many ruinously powerful weapons. Factor in the delicate balance of ecological and cultural systems, and we might as well be children—wide-eyed, insouciant, and untrained—playing with loaded machine guns. It *might* turn out okay, but you wouldn't want to bet the farm on it.

At the same time, we cannot—and we must not try to—halt technological progress. Though human technology creates new hazards for humanity (and the planet), it is absolutely necessary to protect us from natural perils such as hammer blows from space.

Partly for the sake of completeness, but partly, I suspect, for its sheer intellectual fascination, Torres describes some threats that seem more speculative. For instance, we might create mind children—advanced artificial intelligences—that will ultimately turn on us, or perhaps "merely" pursue incomprehensible plans that involve our destruction as a side effect. Or maybe we are living in a *Matrix*-style simulated reality, the creation of unseen masters who don't plan to keep us around forever. Extravagant though these ideas may be, it's surprisingly difficult to rule them out.

For all that, *The End* has a practical purpose. It is a thoughtful, clear, always page-turning manual of big-picture threats. Above all, it asks what our priorities and plans should be if we seriously want to defend our planet and our hard-won level of current civilization. There is no single solution, but nor are we entirely helpless. We can, for a start, investigate the dangers systematically, seriously, and with whatever resources they require—thus founding a science of practical eschatology. We can educate our children to address ongoing problems as genuine and urgent concerns. There is much, perhaps, that we should be doing even now, whether it's constructing planetary defenses against asteroids and comets or developing antidotes to apocalyptic propaganda.

Torres has plenty to say about human priorities in averting global catastrophes, and his advice merits careful thought and judgment. In our ordinary lives, with all the problems that press on us daily, we forget what a dangerous universe we live in, made more so by many of humanity's own efforts. We need to raise consciousness of planetary-level risks, and we need workable plans to address them. Otherwise, there's a prospect of our own extinction event. Please read on, and let us take the issues to heart. Much may depend on it.

Russell Blackford
Newscastle, NSW
Australia

Notes

1. Now more correctly known as the Cretaceous-Paleogene (K-Pg) boundary, but the old terminology remains better known outside the circles of professional geologists.

Preface

· · · · ·

The New Atheism movement has been centrally driven by a single idea: that religion is not merely wrong, but also dangerous. *The End* is the next logical step in this line of reasoning. It merges the insights of New Atheism with the burgeoning field of Existential Risk Studies, most notably associated with the innovative work of the Future of Humanity Institute (Oxford), the Centre for the Study of Existential Risk (Cambridge), and the Future of Life Institute. In bringing these fields into contact, this book shows how faith-based belief in propositions privately revealed to prophets claiming to have special access to the supernatural is intimately connected to the increasingly real prospect that our species either annihilates itself or fails to anticipate (and therefore dodge) threats that could result in our extinction. The fact is that a growing number of technologies capable of turning the apocalyptic fantasies of deranged dogmatists into a reality are populating our world. The stakes, therefore, have never been greater—and neither has the importance of spreading atheism.[*]

Religion is one of many causal factors operating in the world, to be sure. But just as surely, it is a nontrivial factor. Many existential riskologists (as I call them) seem to hold that one can study, and consequently mitigate, the big-picture hazards haunting our future without a deep understanding of the social, cultural, and religious milieus in which world affairs play out and advanced technology is developing. I strongly disagree. Books such as Nick Bostrom and Milan Ćirković's 2008 *Global Catastrophic Risks*—an excellent, albeit academic, introduction to Existential Risk Studies, with papers on a variety of risk scenarios by experts in the relevant fields—tell

---

[*] With the caveat that "conversion" to atheism is motivated by epistemological considerations; see appendix 4.

17

only half the story. In my view, one simply can't understand the formidable task of avoiding disaster without knowing, for example, what the Rapture is, what Sunnis and Shias believe about the Mahdi, the role of Israel in the end-times narrative of dispensationalism, and why the Syrian town of Dabiq is of great importance to a strand of Islamic apocalypticism. To accomplish the goal of "maximiz[ing] the probability of an okay outcome" for humanity, to quote Nick Bostrom, the existential riskologist needs to know about more than science and technology: she must grasp the rudiments of religion, with an eye to how religious beliefs about the future are actively shaping history.[1]

The present book provides this knowledge. Written for both a general audience and academics, it offers a comprehensive snapshot of our collective predicament, in which neoteric technologies and archaic belief systems are colliding with potentially catastrophic consequences.[2] From another angle, *The End* constitutes a kind of "progress report" on the expanding field of *eschatology*, which now has both religious and secular branches. While these branches are distinct, they are not causally independent: they can interact in complicated ways, with one magnifying the dangers posed by the other. As a result, any treatment of this subject must necessarily adopt a *big-picture* approach. *The End* embraces this challenge. It's deeply informed by the sciences yet imbued with the social and historical awareness of the humanities (something lacking in many books on emerging technologies, as well as on atheism). I am a philosopher in training and at heart, in part because philosophy is a horizontal discipline that makes contact with virtually every other field of inquiry. As Wilfrid Sellars famously put it, "The aim of philosophy, abstractly formulated, is to understand how things in the broadest possible sense of the term hang together in the broadest possible sense of the term."[3] This makes the philosopher a *reflective generalist*[4]— precisely the sort of thinker uniquely suited to studying existential risks in the eschatological context of a world full of religious belief.

My hope, therefore, is that the philosophical spirit that suffuses this book will inspire New Atheists and riskologists alike to study *both* branches of contemporary eschatology—to become hybrid thinkers capable of seeing the big picture of civilization's future, without losing sight of the more immediate details. Our ability to understand civilization's unique predicament in the twenty-first century and act accordingly will determine whether we tumble into the eternal grave of extinction or flourish for millions of years to come.

Notes

1.   See Nick Bostrom, "Existential Risks: Analyzing Human Extinction Scenarios and Related Hazards," *Journal of Evolution and Technology* 9, no. 2 (2002).

2.   Indeed, given the breadth of topics covered in the following chapters, it's likely that just about *all* readers will be nonexperts with respect to at least some subjects.

3.   See Wilfrid Sellars, "Philosophy and the Scientific Image of Man," in *Science, Perception, and Reality*, ed. Robert Colodney (Humanities Press/ Ridgeview, 1963), pp. 35–78.

4.   This term comes from Jay Rosenberg, "Wilfrid Sellars," in *Stanford Encyclopedia of Philosophy*, ed. Edward Zalta (Winter 1997 edition), http://plato. stanford.edu/archives/sum1998/entries/sellars/.

1

Looking Forward to the Future

· · · · ·

The Most Important Conversation of Our Age

"Eschatology": "the study of the end of the world."

People in every age have waved their arms in the air and shouted *"The end is near! Prepare for the world to be destroyed!"* Most scholars of the New Testament in Europe and the United States since Albert Schweitzer published *The Quest of the Historical Jesus* (1906) have seen Jesus himself as a failed apocalyptic prophet who expected the imminent end of the world in his own day.[1] This is why Jesus says in Matthew 24, "Truly I tell you, this generation will certainly not pass away until all these things have happened." The Apostle Paul also appears to have anticipated an imminent end. In encouraging men "not [to] look for a wife," for example, Paul says, "What I mean, brothers and sisters, is that the time is short. . . . For this world in its present form is passing away" (1 Corinthians 7).

Since the time of Jesus and Paul, a staggering number of Christians have expected the world to end in their lifetimes, or the foreseeable future. Few Christians are fully aware of how extensive and diverse this tradition is. To pluck just a few examples from the fields of history: Irenaeus, the second-century Church Father who argued, quite influentially, that there should be four Gospels because there are four zones of the world and four principal winds, anticipated Jesus' return in 500 AD.[2] Two centuries later, a bishop named Martin of Tours wrote that "there is no doubt that the Antichrist has already been born. Firmly established already in his early years, he will, after reaching maturity, achieve supreme power." And Martin Luther, of Protestant Reformation fame, expected the world to end before 1600. More recently, Pat Robertson predicted the end in the early 1980s, saying,

"I guarantee you by the end of 1982 there is going to be a judgment on the world."[3] And the retired NASA engineer Edgar Whisenant published a brilliantly incorrect (but nonetheless best-selling) book called *88 Reasons Why the Rapture Will Be in 1988*, which, one might surmise from the title, forecast the Rapture in 1988. Whisenant declared, "Only if the Bible is in error am I wrong."

Perhaps the most famous instance of eschatological embarrassment involved the followers of the American Baptist preacher William Miller, known as the Millerites. These faithful believers anticipated the end on October 22, 1844. As *The Midnight Cry* put it in a final editorial:

> Think for eternity! Thousands may be lulled to sleep by hearing your actions say: "This world is worth my whole energies. The world to come is a vain shadow." O, reverse this practical sermon, *instantly*! Break loose from the world as much as possible. If indispensable duty calls you into the world for a moment, go as a man would run to do a piece of work, in the rain. Run and hasten through it, and let it be known that you leave it with alacrity for something better. Let your actions preach in the clearest tones: "The Lord is coming"—"The Time is short"—"This world passeth away"—"Prepare to meet thy God."[4]

Many Millerites were so convinced that they abandoned their material possessions, gave up their careers, and left their families in preparation for Christ to "cleanse, purify and take possession" of the earth.[5] As the October date approached, some Millerites "failed to plow their fields because the Lord would surely come 'before another winter.'" This belief "grew among others in [the] area so that even if they had planted their fields they felt it would be inconsistent with their faith to take in their crops."[6]

When Jesus failed to emerge from the clouds, many sank into a gloomy despondence. One poor Millerite wrote: "I waited all Tuesday and dear Jesus did not come;—I waited all the forenoon of Wednesday, and was well in body as I ever was, but after 12 o'clock I began to feel faint, and before dark I needed someone to help me up to my chamber, as my natural strength was leaving me very fast, and I lay prostrate for 2 days without any pain—sick with disappointment." Another recorded the experience like this: "The 22nd of October passed, making unspeakably sad the faithful and longing ones; but causing the unbelieving and wicked to rejoice. All was still. . . . Everyone felt lonely, with hardly a desire to speak to anyone. Still in the cold world! No deliverance—the Lord [did] not come!"

More recently, an American evangelist named Harold Camping made international news by declaring that the Rapture would occur on May 21, 2011. He put up more than 3,000 billboards worldwide to announce this event. Like the Millerites of the nineteenth century, some of Camping's followers abandoned their families and jobs to prepare for Jesus' appearance in the sky. One man gave over $140,000 of his own money for Rapture-related advertising. As Camping told Reuters before the big day, "We know without any shadow of a doubt it is going to happen."[7] Meanwhile, atheists planned "rapture parties" for the evening of May 21 and "non-Judgmental Day" activities for May 22. None of these events were cancelled, of course: Camping was wrong.*

As we'll see later on, a large percentage of believers in the contemporary US actively anticipates the imminent end of the world. The influence of these "dispensationalists" on American society and politics has been appreciable: as the historian Paul Boyer puts it in an article linking dispensationalism with the 2003 Iraq War, there's a "shadowy but vital way that belief in biblical prophecy is helping mold grass-roots attitudes toward current US foreign policy." He adds that academics "need to pay more attention to the role of religious belief in American public life, not only in the past, but also today. Without close attention to the prophetic scenario embraced by millions of American citizens, the current political climate in the United States cannot be fully understood."[8] In chapter 13, we'll examine how the powerful Christian Zionist lobby in the US is centrally driven by dispensationalist convictions, according to which the last chapter in God's prewritten narrative of history is crucially dependent upon the existence of a Jewish state in Palestine. The result is an unwavering, dogmatic allegiance to Zionist causes that's fueling some of the more significant conflicts in the world.

* It's worth noting here that the term "cognitive dissonance" was specifically coined in the context of unfulfilled apocalyptic expectations by a few social psychologists who'd infiltrated a UFO cult in the mid-1950s. Those in the cult, called The Seekers, believed that a UFO flown by the "boys upstairs" would save them from a series of world-transforming cataclysms. When this failed to happen, many of the members actually became more convinced of the prophecy's truth, and in fact proselytizing increased. The psychologists argued that this happened because of cognitive dissonance, or the discomfort that occurs when one holds two incompatible beliefs simultaneously (e.g., the apocalypse was supposed to happen and the apocalypse didn't happen). The concept thus had its origins in falsified religious beliefs.

One finds no dearth of eschatological fervor in the Islamic world either, beginning with the prophet Muhammad himself. As Allen Fromherz states in *The Oxford Encyclopedia of the Islamic World*, "some scholars have suggested that Islam was, from the first revelations of Muhammad, almost entirely an apocalyptic movement. . . . Some have even supposed that Muhammad deliberately failed to designate a successor because he predicted that the final judgment would occur after his death."[9] Since Muhammad passed away in the eighth century, many Muslims have claimed to be the Mahdi, a messianic figure prophesied in the *hadith* (a collection of Muhammad's deeds and sayings) to unite the Muslim world (*ummah*) and, with the help of Jesus (or *Isa*, who will descend to Earth from heaven), defeat the *Dajjal* (the Muslim Antichrist) in a battle of monumental proportions.

A rather dramatic instance of Islamic eschatology being taken seriously in recent times occurred in 1979, when an apocalyptic group stormed the Grand Mosque in Mecca, the holy city of Muhammad's birth. The Grand Mosque is the biggest worship center in the Muslim world, and it's built around the holiest site in Islam, the Kaaba, toward which Muslims face when they pray. While 100,000 worshipers were inside the mosque on November 20, some 500 insurgents gained control of the building.[10] These men claimed to have the Mahdi among them: a man named Mohammed Abdullah al-Qahtani. Saudi forces attempted to take back the mosque but repeatedly failed. It took a whole two weeks of fighting for the insurgents to finally surrender, probably *after* the "Mahdi" was killed (thereby proving their eschatological claims wrong). Most of the rebels who survived this incident—more than sixty—were publicly beheaded in 1980, a form of execution still practiced in modern-day Saudi Arabia.

Many others have come forward as the Mahdi. Dia Abdul Zahra Kadim, for example, the leader of the Shia extremist group Soldiers of Heaven, believed he was the Mahdi. He was killed in 2007 fighting the coalition forces in Iraq. That same year, an influential Iranian cleric (who's notable for supporting church and state separation) was sentenced to eleven years in prison "for allegedly—among other things—claiming he was the Mahdi." As of 2013, over 3,000 people claiming to be this messianic figure were, in fact, living in Iranian prisons.[11] And the leader of arguably the most powerful and well-funded terrorist organization in history, the Islamic State, has been reported "to be trying to style himself as the Mahdi" (see chapter 13 for contrary opinions).[12] This Sunni group is explicitly motivated by apocalyptic visions of the end, and it believes we're on the verge of an epic battle—essentially, Armageddon—that's set to occur in a small town in northern Syria. Chapter 13 will examine this topic in much greater detail.

Zooming out a little more, one finds a significant number of terrorist groups around the world today driven by delusions of history's imminent conclusion. Boko Haram, for example, is a radical Islamic group based in Nigeria whose name translates as "Western education is forbidden." In 2015, they officially pledged allegiance to the Islamic State, thereby embracing its eschatological worldview of impending catastrophe. Further east, a millennial cult known as the Eastern Lightning preaches that a woman in central China is the reincarnation of Christ and "that the righteous are engaged in an apocalyptic struggle against China's Communist Party— which they refer to as the 'great red dragon.'"[13] This group is named after the prophetic verse of Matthew 24:27, which states that "for as lightning that comes from the east is visible even in the west, so will the coming of the Son of Man," and as of 2012 it was estimated to have around a million followers.[14] Finally, here in the US, the Christian Identity movement is a terrorist group that adheres to an ideology whereby "the end of the current world order is close, that they need to take some active role in promoting this event, and that this apocalyptic event is an imperative to be furthered with the use of violence."[15]

Thus, from Europe to the United States to the Middle East and China— across space and time, history and geography—end times, or *millenarian*, thinking is ubiquitous. *One simply can't understand world affairs without understanding how the major religions think it will end.* Yet every single end-times enthusiast to date has been wrong about his or her predictions. Some observers take this to imply that *anyone* claiming the world may soon come to an end is probably wrong (and perhaps crazy). This tendency (for reasons explored in appendix 4) ought to be resisted, at least in a qualified way. The fact is that a thousand people having cried "Wolf!" without an attack doesn't mean that a vicious canine *isn't* creeping up behind us. What ultimately matters are the *reasons* behind one's cries, the *arguments* one has for arm-waving and shouting kooky things like "The end is near!"

This leads to an important distinction within the contemporary field of eschatology. Ever since the ancient Persians (in particular, the prophet Zoroaster) basically invented eschatology, this field has consisted of only a single branch: *religious eschatology*.[16] By the middle of the twentieth century, though, it sprouted a second: *secular eschatology*. There are several notable differences between these branches. First, whereas religious narratives typically involve miraculous events and supernatural agents, secular eschatologies are thoroughly "naturalistic." There's nothing awaiting us on the other side of the grave, according to naturalism (also called materialism, and physicalism), no eternal paradise for "the elect" to inhabit once this world wastes away. So the stakes are high—perhaps *far higher* for the secular eschatologist, since this life is our one and only shot.

Along these lines, religious eschatologies include, quintessentially, some portion of humanity surviving a catastrophic transformation of the cosmos as the forces of Good overcome Evil through a series of climactic confrontations. This world soaked with sin and suffering is thus destroyed and along with it all those who failed to convert to the One True Religion (whatever it happens to be). In contrast, secular eschatologies concern, quintessentially, a special category of tragedies known as *existential risks*. We'll examine these in detail below; for now, we can say that an existential risk involves either no one surviving or some portion of humanity surviving but coming to inhabit a postcatastrophe world marked by severe and irreversible privation. Examples of possible existential risks include a nuclear war between world powers, a terrorist attack using designer pathogens, and a supervolcanic eruption.

The most crucial difference between the religious and secular branches of eschatology, though, pertains to their radically different *epistemological foundations*. Whereas the end-times stories of religion are based on faith in prophecies acquired through revelation, those of the secular branch are founded on empirical evidence acquired through observation. It's this fundamental philosophical difference that accounts for why we should take the cries of "Wolf!" made by secular riskologists (*very*, in some cases) seriously, while dismissing the pious prophets' tales of raptures, resurrections, and the return of supernatural entities as unjustified nonsense. Indeed, it's epistemology that makes the difference between *being alarmed* and *being an alarmist*, two attitudes often confused in public debates about big-picture risks like global warming.* As we'll explore in chapters 12 and 13, a large portion of humanity is far more interested in

---

* Yet another difference is that the apocalypses of religion are supposed to happen. That is to say, religious eschatology is teleological in a way that secular eschatology isn't.

the eschatological narratives of religion than the doomsday scenarios put forth by evidence-minded scientists. The result is that, given the superior epistemological status of secular eschatology, *most people are fixated on the wrong apocalypses*.

This is quite terrifying, especially since the best current estimates suggest that the probability of *Homo sapiens* following the dodo into extinction is greater than it's ever been before in our history (in part because people are looking the wrong direction!). The fact is that we live in a qualitatively different epoch in human history, one marked by a steadily increasing number of existential risk scenarios (see chapter 14). And while we have a good track record of avoiding the background risks that have always haunted us in nature—you wouldn't be reading this if our track record were bad—we have hardly any record of dodging the threats facing us in the present and coming centuries. We're charging into uncharted territory, rushing from the twilight into darkness with hardly a constellation above to guide us.

This makes the topic of existential risks quite possibly the most important that one could study. Indeed, I began writing this book because I don't believe there's any subject of greater *ultimate value*: everything we care about in the world, in this great experiment called civilization, depends on us preventing an existential catastrophe. So do all future humans—potentially billions and billions of people living lives that could be even *more* valuable, relative to some criteria of self-actualization and prosperity, than ours. There is, in fact, a real possibility that our children could live indefinitely long lives, become superintelligent uploaded minds, and even colonize the universe, traversing the great expanses of space at near the cosmic speed limit of light. (If this sounds like quixotic speculation, imagine explaining a smartphone, cochlear implant, or jet engine to a Neanderthal.) Thus, given how high the stakes are, not to mention that end-times thinking has had a pervasive (and perverse) effect on world affairs, the field of eschatology constitutes arguably the most urgent and imperative conversation of the present day.

Big-Picture Hazards: Definitions

Let's take a closer look at eschatology's secular offshoot for a moment. This section will be one of the more technical parts of the book, but the challenge will be worth it, given that existential risks are the heart and soul of secular eschatology.

What exactly is an existential risk? Historically speaking, the *concept* dates back at least to the middle of the last century. After the US dropped two atomic bombs on the unsuspecting folks of the Japanese archipelago, the possibility of a secular apocalypse became a major topic of conversation among academics and the public alike.* The word "existential risk" is also not new, although it doesn't quite date back to 1945. A 1988 paper by Peter Frey, for example, uses it in nearly its contemporary sense; foreshadowing our discussion below, Frey writes that "existential risks may be either natural or man-made in origin. More and more of the supposedly 'natural' existential risks can, on closer analysis, be identified as man-made."

Nonetheless, it's the work of Nick Bostrom, an Oxford University philosopher and director of the Future of Humanity Institute (FHI), that's been largely responsible for the popularization of the term over the past decade. For the purposes of this book, we'll define a risk as the probability of an undesirable event multiplied by its consequences. This is the standard definition in the actuarial sciences. But it can be made more precise by analyzing the consequences of a risk in terms of intensity (how severe the effects are) and scope (how many people are affected), as Bostrom does in his influential paper "Existential Risks: Analyzing Human Extinction Scenarios" (2002). While Bostrom leaves his analysis at this, we can take a step further and distinguish between the *spatial* and *temporal* scopes of a risk's consequences (figure A). This distinction is motivated by the fact that risks can have varying spatiotemporal effects. For example, a population of 1 billion could be impacted by a one-off event that happens in a single instant. Or, alternatively, the same population size could be impacted by an ongoing phenomenon that affects only a hundred people at a time but recurs over many generations until 1 billion people have been affected. Our responses to, and strategies for avoiding, these two scenarios might be completely different, and this is why an adequate account of risks ought to enable the riskologist to distinguish between them.[17]

The analysis of figure A yields the typology of risks shown in figure B. Whereas the analysis is a dissection of the risk concept, the typology is a classification scheme of risk types based on different combinations of their properties. In accordance with figure A, figure B contains three distinct

---

* John Leslie's **The End of the World**, first published in 1996, provides an interesting and comprehensive discussion of existential risks, although using different terminology.

Figure A. Dissecting the Concept of a Risk

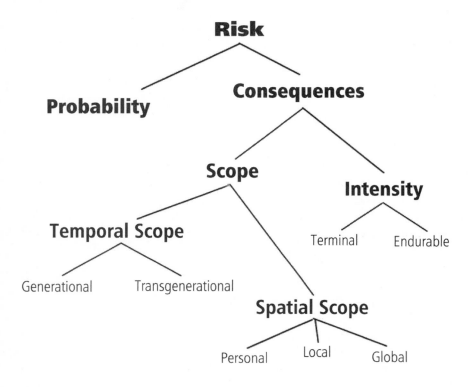

axes, one corresponding to the temporal scope, one to the spatial scope, and one to the intensity of a risk. The first is divided (somewhat arbitrarily) into generational and transgenerational categories;[18] the second into personal, local, and global; and the third into endurable and terminal.

To fill in this typology with a few concrete examples: a physics disaster that destroys the universe would be a red-dot event; an oppressive government gaining global control could be a black-dot event; aging, which affects everyone everywhere with death but doesn't (necessarily) result in the termination of our species, corresponds to the orange dot;[19] the disappearance of the Republic of Maldives (an island nation) due to rising sea levels corresponds to the green dot; a volcanic eruption might correspond to the purple dot; Superstorm Sandy, which slammed New York City in 2012, corresponds to the grey dot; a survivable germline mutation

Figure B. A Working Typology of Risks

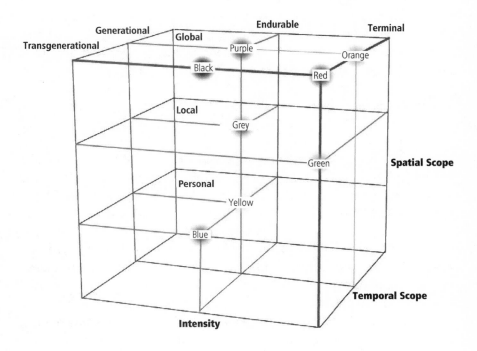

passed down from one generation to the next, corresponds to the blue dot;[20] and a survivable case of skin cancer corresponds to the yellow dot. Figure B has a place for virtually any kind of risk one might want to categorize.

With this in mind, we can now begin to piece together a definition of existential risks. The intuition behind this concept is that it's supposed to latch on to the *worst possible scenarios* for either our current population or our descendants. Clearly, red-dot events are bad—*really* bad: they result in the total annihilation of our species. So it seems that existential risks should at least include catastrophes of this sort. But do red-dot events exhaust the definition of this term? The Oxford philosopher Derek Parfit seems to think so.* In a famous passage from his book *Reasons and Persons*, he argues that the difference between 99% and 100% of humanity perishing

---

* Although Parfit doesn't explicitly use the term "existential risks."

is *far greater* than the difference between 1% and 99% of humanity kicking the bucket.[21] (Most people would intuitively say the opposite, given that the first case only involves an extra 1% of humanity dying out, whereas the latter involves a whopping 98%.) Why does Parfit claim this? Because the 100% scenario affects not only all humans alive at the time but all future possible humans who could potentially live lives even more valuable than ours. Red-dot catastrophes close off the possibility of future civilization, and this is what makes them *profoundly worse* than any other type of risk, even those resulting in the death of nearly everyone. Such is Parfit's reasoning.

But is it correct? One might wonder about survivable catastrophes that severely compromise human prosperity and cannot be recovered from, *ever*. Imagine, for example, a situation in which future people are forced to live under the tyrannical rule of a sadistic alien regime until the sun sterilizes the earth in several billion years.[22] What if life in this state were worse than it is for those living in North Korea, and conditions stayed this way until the end of time? Is this scenario really better, as Parfit seems to imply, than our species going extinct? I would answer in the negative, and many riskologists would agree. Putting this in terms of figure B, there are some black-dot events (i.e., global-*endurable*-transgenerational catastrophes) that are *equally as bad* as those of the red-dot variety. If one were forced to choose between enduring a black-dot catastrophe of this sort *or* a red-dot disaster, it would be hard to decide which to pick: surviving a black-dot catastrophe that's both severe and irreversible appears *no less dreadful* than total annihilation.

We're now in a position to define "existential risks" with conceptual precision. Using figure B, we can stipulate that an existential risk is equivalent to (i) *all* red-dot catastrophes, and (ii) *any* black-dot catastrophe that significantly compromises human well-being.* It follows that a scenario need not be *terminal* to count as an existential risk; the only features shared

---

* Notice here that the category of "terminal" is binary. That is to say, something is either dead or alive, extant or extinct. This fits our ordinary way of speaking. On the other hand, the category of "endurable" is spectral, as there exists a range of endurable events, from the imperceptible to the world-changing. Thus, within the second category, black-dot existential risks are those with extremely severe, world-changing consequences, where these consequences are deemed morally undesirable.

by all existential risks are their global and transgenerational scopes. These are their two *necessary conditions*.

Notice that existential catastrophes are completely unique among all other risk types. They are unlike every other threat our species has ever before encountered. The reason is that they are, by definition, *historically singular events* for a given species.[23] As a result, existential risks present a special challenge for humanity, since our usual method of avoiding dangers in the future is to glance over our shoulders and take a look at the past. But we can't do this with existential risks: we can't learn from our mistakes the way we do with other kinds of disasters. If even a *single* existential risk were to occur, the game would be over, and we'd have lost. It follows that we must rely entirely on *anticipation* rather than *retrospection* to dodge such threats; we must be proactive rather than reactive.[24] There are, simply put, no second chances when it comes to existential risks.

Furthermore, just because certain existential risks have never before occurred doesn't necessarily mean they're improbable. Why? Because the only universes in which we could possibly find ourselves are ones in which such an event has never before happened. That is to say, some events are *unobservable* because they entail the elimination of observers. We can talk about a red-dot catastrophe happening tomorrow, but not about one having happened yesterday. Consequently, we may be significantly underestimating the likelihood of extinction.* As Milan Ćirković and Bostrom make the point, "We are bound to find ourselves in one of those places and belonging to one of those intelligent species which have not yet been destroyed, *whether planet or species-destroying disasters are common or rare*" (italics added).[25] This is called the *observation selection effect*. It entails that we can't project a record of past successes onto the screen of the future—having survived this far doesn't mean we're in any way *safe* from total destruction.[26] In fact, taking a cosmic perspective, some estimates based on the Drake Equation (used to assess the number of technological civilizations that might exist) suggest that the universe should be teeming with life. Yet we hear neither a whisper nor a shout for cosmic companionship in the darkness of space. The universe is utterly silent—and eerily so. Perhaps the destruction of intelligent life is, therefore, highly likely, and we're the outliers. Perhaps the probability of extinction is extremely high, and we've just been lucky.

* A point to which we'll return in chapter 10.

Having said this, existential risks aren't the only type of risk worth worrying about. There's a broader category of troublesome events worthy of our attention that I'll call *big-picture hazards*. This term denotes any catastrophe whose consequences are global in spatial scope, whatever other properties it might possess. In other words, globality is the single *essential property* of this category. With respect to figure B, big-picture hazards include all purple-dot, orange-dot, red-dot, and black-dot disasters. This makes it nearly synonymous with the more commonly used term "global catastrophic risks." But the semantic overlap isn't total, since global catastrophic risks are typically defined in terms of risk typologies that fail to make the same distinctions as figure B (see appendix 1).

Since the realization of *any* big-picture hazard would result in significant harm to civilization, much of this book will examine not just existential risks but nonexistential risks as well. In fact, the probability that one or more global catastrophes will devastate civilization this century may be *far greater* than the probability of an existential disaster. Indeed, with respect to most types of risks, the less severe the consequences, the higher the probability.* It may even be the case, as we'll discuss later, that successfully defending against an existential catastrophe will have the unfortunate effect of *increasing* the likelihood of one or more big-picture hazards. In other words, there may be a tradeoff between survival and suffering such that the more secure our survival, the more likely we are to suffer from a series of unprecedented tragedies, and *vice versa*. Considerations like these make the more general class of big-picture hazards worth examining, not just the special case of existential risks.

Causes of Big-Picture Hazards

One can't defend against existential risks without understanding their causes. For the past 2 million years or so, since the first human (*Homo*) species emerged from the grassy savanna of East Africa, our genus has been haunted by a handful of improbable hazards. The list includes, most notably, asteroid and comet impacts, supervolcanic eruptions, and

---

* For example, it's more likely that half of humanity dies out in a global pandemic than all of us, just as it's more likely that 1 million people perish in a nuclear exchange than 5 billion. There are some risks, though, where this relation doesn't hold, as we'll explore in the final chapter of the book.

pandemics. Aside from these natural events, there wasn't much else that could have terminated our lineage.

The vast majority of existential risks today, though, are anthropogenic. They arise from human activities rather than nature. As we'll see in chapter 14, it's difficult to specify an exact number of anthropogenic threats this century, given that some are rather speculative, but it wouldn't be misleading to say that there are about twenty human-created threats in total—and the number keeps growing.[27] An important subcategory of anthropogenic risks is what we'll call *technogenic risks*, or existential scenarios that arise from the misuse and abuse of advanced technologies. Whereas global warming and biodiversity loss are anthropogenic risks, they aren't technogenic, at least not in the sense that a bioterrorism attack or a war involving nanotechnology are. By far the greatest cause for apocalyptic anxiety these days comes from risks that are technogenic in origin.

Let's explore this phenomenon for a moment, since it will be a recurring theme in this book. Technogenic risks are the byproduct of a particular set of properties had by advanced technologies. To begin, most or perhaps all advanced technologies are *dual-use* in nature. This means that the very same artifacts, techniques, theories, research, information, knowledge, and so on can be used for both morally good and morally bad ends; their space of possible application includes benevolent as well as nefarious uses. Examples of dual usability include the centrifuges that enrich uranium for nuclear power plants, as they can also be used to enrich uranium for nuclear weapons. Similarly, the very same knowledge that could lead to a cure for Ebola could also be exploited to create a designer pathogen. What's absolutely crucial to understand about dual usability is that the good and the bad are a package deal: one comes with the other, and *to eliminate either is to eliminate both*.

But dual usability wouldn't be an occasion for eschatological concern if it weren't for another property of advanced technology: its extraordinary *power*, which in some cases is growing at something like an exponential, or possibly even an exponentially exponential, rate.[28] As a result, the human ability to manipulate and rearrange the world in different ways is expanding rapidly. We can now design genomes on the computer, destroy entire cities with a single bomb, and even create (in the foreseeable future) microscopic robots capable of "eating" the entire biosphere. The fact is that we aren't children playing with matches anymore; we're children messing around with flamethrowers.

But not only are advanced technologies becoming more powerful, some are also becoming increasingly *accessible*. This applies most notably to the fields of biotechnology, synthetic biology, and nanotechnology. The accessibility of such fields, I would argue, can be measured along four distinct axes: intelligence (raw brainpower), knowledge (know-that), skills (know-how), and equipment (the material means). That is to say, the intellectual capacity needed to use certain technologies to radically modify the world is quickly decreasing. One need not be an "evil genius" to bring about large-scale catastrophes these days—and this will be even more true in the future. As the AI theorist Eliezer Yudkowsky puts it, the result is "Moore's Law of Mad Science," which states that "every eighteen months, the minimum IQ necessary to destroy the world drops by one point."*

Second, it's undeniable that one needs less of an education to perform the same manipulations as the experts. The terrorist of the future might not need a master's, or even a bachelor's, degree in microbiology to create a pathogen capable of causing widespread misery and death. An autodidact could spend just a few weeks watching lectures on YouTube, reading textbooks downloaded from libgen.org, and acquiring the genomes of viruses like polio and smallpox from publicly accessible, online databases (see chapter 3). Following Yudkowsky, then, we might propose another version of Moore's Law of Mad Science, this one focusing not on IQ but on the amount of knowledge needed to inflict harm on society.

Third, it appears to be the case, generally speaking, that fewer skills are required to perform such manipulations. Just because one has the theory doesn't mean that one can put it into practice. But the evidence suggests that the practical abilities necessary to conduct certain experiments are on the decline. As the Princeton scholar Christopher Chyba observes, many "laboratory processes have become more automated and black-boxed so that less and less tacit knowledge [i.e., practical know-how] is needed to employ the technologies."[29] This trend, whereby the skill gradient that separates elite practitioners and nonexperts is incrementally eroded, is called *de-skilling*.[30] It yields yet another version of Moore's Law of Mad Science, this one pertaining to skill rather than IQ or knowledge.

Finally, the instruments and materials needed to set up, say, a genetics laboratory behind the walls of one's apartment are becoming evermore

---

* This isn't intended to be mathematically precise. It's merely supposed to gesture at a real trend that's been unfolding the past few decades.

affordable. All it takes to build a home laboratory today is a few thousand dollars' worth of equipment, which one can purchase from online companies like Open PCR (openpcr.org). Meanwhile, DNA sequences can be ordered from companies like Macrogen (dna.macrogen.com). As we'll discover in chapter 3, it's surprisingly, even *consternatingly*, easy for scientists and nonscientists alike to acquire strands of DNA from some of the world's most deadly pathogens. The result is a fourth version of Moore's Law of Mad Science, this one focusing on the cost of the equipment and instrumentation necessary to set up a fully functional home laboratory.

In sum, advanced technology is dually usable in nature, it's becoming increasingly powerful, and in some cases it's placing this power in the hands of single individuals. This is worrisome because there are many people in the world who harbor a death wish for our species, who fantasize about the collapse of civilization. But the situation is worse than this. The fact is that one need not be an evil misanthrope with deathist daydreams to bring about a catastrophe of unprecedented proportions. This leads to an important distinction within the category of "bad applications," between what Sir Martin Rees calls *error* and *terror*.[31] The difference here concerns the intentions behind the destructive act, not the degree to which the act is destructive: in other words, an error could cause our extinction no less than terror. As one might surmise, the former category includes events resulting from negligence, mistakes, accidents, technical failures, momentary lapses of reason, and clumsy fingers, whereas the latter includes events caused on purpose. Following tacit convention, I'll define "terror" as being a broad category that subsumes acts of terrorism as well as phenomena like wars between states.

As chapter 14 elaborates in detail, our best chance of avoiding an existential catastrophe (without giving up on civilization entirely) is to neutralize the categories of error and terror. There are many ways to do this, it turns out, but most require one or more improbable leaps of human moral progress and rationality. If we fail at this task, though, an existential risk may be all but guaranteed in the coming centuries.

Speculation and Predictability

Secular eschatology could be located within the field of Future Studies (or Futurology). Future Studies is concerned with the "three Ps": possibility (what could happen), probability (what's likely to happen), and preferability (what we want to happen). There's a tendency among some scholars to dismiss this field because of the perceived difficulty of saying anything substantive about the future: who knows what the world in 10, 20, or 100 years might be like? Just recall what people in the 1950s thought about the year 2000, or the claims made by artificial intelligence researchers in the 1960s about how the revolutionary arrival of ultraintelligent machines is right around the corner. Trying to tell true stories in future tense is, the argument goes, a waste of time, because it can't be done with any degree of accuracy.

This criticism is only half correct. The first thing to note is that *every field* of inquiry that makes claims about the future has, by definition, a futurological component. In the hard sciences, one of the primary aims of a theory or hypothesis is to *predict* what will happen if a particular set of conditions is satisfied. According to Einstein's theory of relativity, the astronauts orbiting the earth in the International Space Station (ISS) should literally age *slower* than those of us back home. The reason is that velocity slows time down—in this instance, not from the subjective perspective of those on the ISS, but relative to you and me on the earth. This phenomenon is called "time dilation." It entails that if you had a twin that traveled to Pluto and back, he or she would return *younger* than you. As it happens, gravity also slows down time, meaning that the closer to the earth you are, the slower time will pass (relative to those further up). It follows that your scalp will literally age *faster* than your toes, since your toes spend more time closer to the earth.

These are predictions made by Einstein's theory. They are specific claims about what we can expect to happen in the future. And experimental evidence shows that they're extremely accurate. For example, the clocks on the ISS *really do* run slower than those on Earth. And experiments have detected differences in the passing of time "due to a change in height near Earth's surface of less than 1 meter."[32] The point is that a central feature of theories in physics and related fields is futurological in nature. In this sense, Futurology is a *horizontal* field that spans a wide range of other disciplines. Wherever one finds predictive hypotheses, one also finds Futurology.

Second, predictability forms a spectrum. Clustered at one end are phenomena that can be anticipated with a high degree of accuracy—in some cases with *virtual certainty*. For example, in 2014, scientists managed to land a spacecraft called Rosetta on the comet 67P/Churyumov-Gerasimenko. This was, as numerous articles put it, like landing a washing machine on a speeding bullet some ~3.9 billion miles away.[33] In order to do this, scientists needed to know *exactly* where the comet was going to be in the future. Combining some observational facts with the known laws of physics, they were able to make an exceptionally good prediction of the future. Other examples of tractable phenomena include lunar and solar eclipses, how much energy two atoms of hydrogen and one of oxygen will release when combined, what will happen in the event that a nuclear chain reaction occurs, and when the next blood moon will take place (an event that, incidentally, has significant eschatological import for some Christians, due to prophetic verses like Joel 2:30–31 and Revelation 6:12).

Clustered at the other end of the spectrum are events that completely elude prediction. For example, it's virtually impossible to say what the next big communication technology will be. It wasn't more than twenty years ago that people had *not even an inkling* that smartphones would be invented, becoming as common as the pockets in which they reside. Similarly, it's virtually impossible to predict who will win the US presidential election of 2060, or the exact temperature in Johannesburg, South Africa, at 12:01pm five weeks from today, or when a major breakthrough in life extension technology will occur.

The universe is *deterministic*, meaning that *if* one knew every detail about the cosmos and every law of nature, *then* one would be able to predict the future with absolute precision.* But we don't know every detail and might not know every law. It follows that the reason events like those above are hidden from view concerns the limits of our understanding: we just don't know enough (not even close!) to trace the causal structure of the universe up to the year 2060 to see who wins the US election. It follows that *forecasting the future is both entirely doable and hopelessly impossible.*

The big-picture hazards discussed below are scattered across this spectrum of predictability. Some can be assigned robust probabilities based on extensive records of objective historical evidence. This is the case for asteroid/comet impacts and supervolcanoes, for example. While there's a good record of past pandemics, too many factors affect the emergence of infectious disease to accurately extrapolate from this record. When one can't assign a probability based on objective evidence, one is forced into

---

* I'm simplifying quite a bit here. Whether the universe is deterministic isn't a settled matter. For an excellent discussion of the issue by the theoretical physicist Sean Carroll, see, "On Determinism," Preposterous Universe, http://www.preposterousuniverse.com/blog/2011/12/05/on-determinism/.

the realm of *speculation*. Here the probability assignments are "subjective" rather than "objective." They are based on hunches rather than hard data. This doesn't in any way mean, though, that such probabilities must be arbitrary or haphazard. Subjective estimates can still be informed and constrained by the best available evidence; they can still be built on truth-preserving inferences from empirical observations of the world. This is precisely the case with what's probably the most speculative scenario considered in this book: that we're living in a computer simulation, and it gets shut down. Although this catastrophe may sound outrageous to the untrained ear, the reasoning behind it is based on a highly plausible assumption about the nature of consciousness, a weak version of the Principle of Indifference, and some straightforward observations of current technological trends.

It follows that speculation and probability are orthogonal axes of futurological conjecture: the probability of a prediction is *independent of* its speculativeness. For example, one could assign an objective probability of 5% to a risk X occurring, while assigning a subjective probability of 95% to a risk Y. Indeed, some of the most qualified riskologists in the world are more worried about certain *speculative* scenarios, such as a superintelligence taking over the world, than thoroughly concrete possibilities, such as an asteroid smashing into our planet. The latter is a well-understood threat, but it's not seen by many (or any) riskologists as being more significant than the former.

There are reasons to be wary of subjective probabilities, though, since speculation opens up space for *cognitive biases* to creep in and muddle our thinking. In dealing with speculative scenarios, therefore, we must constantly guard against bad habits of thought such as the hindsight bias, anchoring effects, the conjunction fallacy, overconfidence, the availability heuristic, observation selection effects, poverty of imagination arguments, and the gambler's fallacy, to name just a few.[34] Such temptations of the mind could easily lead us to conclude that a risk X is unlikely when it's not, or that Y isn't likely when it is.

The gist of this section is that one shouldn't dismiss a scenario simply because it's speculative. Some of the risk narratives examined below are highly speculative, yet their foundations are ultimately scientific, and this is what matters. As appendix 4 explores in detail, outrageous claims are only outrageous to believe if they lack a sufficiently good evidential support base.

Notes

1.   For more, see Bart Ehrman, "Who Can Still Be a Christian," *Christianity in Antiquity: The Bart Ehrman Blog*, November 10, 2013, http://ehrmanblog.org/can-still-christian/.

2.   See Bart Ehrman, *Misquoting Jesus: The Story Behind Who Changed the Bible and Why* (HarperOne, 2007), p. 35.

3.   More recently, in 2014, Robertson claimed on the *700 Club* that ISIS is fulfilling biblical prophecy.

4.   This excerpt also appears in Leon Festinger, Henry Riecken, and Stanley Schachter, *When Prophecy Fails* (Pinter & Martin Ltd, 1956), p. 23.

5.   Quoted from Charles Farhadian, *Introducing World Religions: A Christian Engagement* (Baker Academic, 2015), p. 482.

6.   See Francis Nichol, *The Midnight Cry: A Defense of the Character and Conduct of William Miller and the Millerites, Who Mistakenly Believed that the Second Coming of Christ Would Take Place in the Year 1844* (TEACH Services, Inc., 2000, originally published in 1944), p. 213.

7.   See "UPDATE 1—Predictor of May 21 Doomsday to Watch It on TV," *Reuters*, May 19, 2011.

8.   See Paul Boyer, "When US Foreign Policy Meets Biblical Prophecy," *Alternet*, Feburary 19, 2003, http://www.alternet.org/story/15221/when_u.s._foreign_policy_meets_biblical_prophecy.

9.   From Allen Fromherz, "Final, Judgment," in *The Oxford Encyclopedia of the Islamic World*, ed. John Esposito (February 2009 edition), http://www.oxfordislamicstudies.com/article/opr/t236/e1107. A leading expert on Muslim eschatology, David Cook, says something similar in *Contemporary Muslim Apocalyptic Literature* (Syracuse University Press, 2008): "Islam probably began as an apocalyptic movement, and it has continued to have a strong apocalyptic and messianic character throughout its history, a character that has manifested itself in literature as well as in periodic social explosions" (p. 1).

10.   The "100,000" number comes from Steve Inskeep's interview with Yaroslav Trofimov; see "1979: Remembering 'The Siege of Mecca,'" *NPR*, August 20, 2009, http://www.npr.org/templates/story/story.php?storyId=112051155.

11.   See "Iran's Multiplicity of Messiahs: You're a Fake," *Economist*, April 27, 2013, http://www.economist.com/news/middle-east-and-africa/21576700-authorities-think-too-many-people-are-claiming-be-mahdi-youre.

12. See Giles Fraser, "To Islamic State, Dabiq Is Important—But It's Not the End of the World," *Guardian*, October 10, 2014, http://www.theguardian.com/commentisfree/belief/2014/oct/10/islamic-state-dabiq-important-not-end-of-the-world.

13. See Tim Hume, "'Eastern Lightning': The Banned Religious Group That Has China Worried," *CNN*, February 3, 2015, http://edition.cnn.com/2014/06/06/world/asia/china-eastern-lightning-killing/index.html.

14. See John Hannon, "China Cracking Down on Doomsday Group," *Los Angeles Times*, December 17, 2012, http://articles.latimes.com/2012/dec/17/world/la-fg-china-doomsday-arrests-20121218.

15. See Charles Ferguson and William Potter, *The Four Faces of Nuclear Terrorism* (Routledge, 2005), p. 18.

16. Or so some scholars, such as Norman Cohn, have argued.

17. This section draws heavily, and in some cases *ad verbum*, from Phil Torres, "A New Typology of Risks," IEET, February 5, 2015, http://ieet.org/index.php/IEET/more/torres20150205, and Phil Torres, "Problems with Defining an Existential Risk," IEET, January 21, 2015, http://ieet.org/index.php/IEET/more/torres20150121.

18. The distinction between generational and transgenerational effects could easily be elaborated (perhaps indefinitely) to achieve greater precision. For example, we could distinguish between generational, transgenerational, and pangenerational, where the latter covers all eternity, transgenerational covers anything that affects more than one generation, and generational covers risks within a single generation of humans. In my opinion, the generational/transgenerational distinction satisfices for the purposes of this book, which is why I use it in figure B.

19. This is yet another problem for Bostrom's typology; see appendix 1 for more.

20. Note that Nick Bostrom's typology cannot accommodate this risk; see appendix 1 for more.

21. I'm rephrasing this from the original to make it less confusing (Parfit talks about "peace" instead of "1%").

22. According to one study. See Dennis Overbye, "Kissing the Earth Goodbye in about 7.59 Billion Years," *New York Times*, March 11, 2008, http://www.nytimes.com/2008/03/11/science/space/11earth.html.

23. As Milan Ćirković writes, "One important selection effect in the study of [global catastrophic risks] arises from *the breakdown of the temporal symmetry*

between past and future catastrophes when our existence at the present epoch and the necessary conditions for it are taken into account. In particular, some of the predictions derived from past records are unreliable due to observation selection [discussed in the body text momentarily], thus introducing an essential qualification to the general and often uncritically accepted gradualist principle that 'the past is a key to the future.'" See Milan Ćirković, "Observation Selection Effects and Global Catastrophic Risks," in *Global Catastrophic Risks*, ed. Nick Bostrom and Milan Ćirković (Oxford University Press, 2008), p. 121.

24. See Nick Bostrom, "Existential Risks: Analyzing Human Extinction Scenarios and Related Hazards," *Journal of Evolution and Technology* 9, no. 1 (2002).

25. See Nick Bostrom and Milan Ćirković, "Introduction," in *Global Catastrophic Risks*, ed. Nick Bostrom and Milan Ćirković (Oxford University Press, 2008), p. 10.

26. As Ćirković puts it, to repeat footnote 23, it's a fallacy to "claim that we should not worry too much about existential disasters, since none has happened in the last thousand or even million years ... predictions derived from past records are unreliable due to observation selection, thus introducing an essential qualification to the general and often uncritically accepted gradualist principle that 'the past is a key to the future.'" See Milan Ćirković, "Observation Selection Effects and Global Catastrophic Risks," in *Global Catastrophic Risks*, ed. Nick Bostrom and Milan Ćirković (Oxford University Press, 2008), p. 121.

27. I get this number from a rough count of the anthropogenic risks listed in Nick Bostrom, "Existential Risks: Analyzing Human Extinction Scenarios and Related Hazards," *Journal of Evolution and Technology* 9, no. 1 (2002).

28. As Kurzweil writes in *The Singularity Is Near: When Humans Transcend Biology* (Penguin Group, 2005), "For information technologies, there is a second level of exponential growth: that is, exponential growth in the rate of exponential growth (the exponent)" (p. 25).

29. See Christopher Chyba, "Biotechnology and the Challenge to Arms Control," Arms Control Association, https://www.armscontrol.org/act/2006_10/BioTechFeature.

30. I'm paraphrasing here from Guatam Mukunda, Kenneth Oye, and Scott Mohr, "What Rough Beast? Synthetic Biology, Uncertainty, and the Future of Biosecurity," *Politics and the Life Sciences* 28, no. 2 (2009).

31. See Martin Rees, *Our Final Hour: A Scientist's Warning* (Basic Books, 2003).

32.   See C. W. Chou, D. B. Hume, T. Rosenband, and D. J. Wineland, "Optical Clocks and Relativity," *Science* 329 (2010): pp. 1630–1633, http://tf.boulder.nist.gov/general/pdf/2447.pdf.

33.   To be precise, this is the *total distance* traveled by Rosetta over a course of ten years.

34.   See Eliezer Yudkowsky, "Cognitive Biases Potentially Affecting Judgement of Global Risks," in *Global Catastrophic Risks*, ed. Nick Bostrom and Milan Ćirković (Oxford University Press, 2008).

2

Fire and Ice

• • • • •

The Doomsday Clock Is Ticking

One of the most significant secular risks to our future comes from nuclear weapons. Although humanity has managed to avoid a nuclear holocaust so far, we've come shockingly close to disaster many times since the Atomic Age commenced in 1945. For example, President John F. Kennedy estimated the probability of an all-out exchange during the Cuban Missile Crisis to be "between 1 in 3 and even." In other words, we only needed to replay this debacle two or three times for a nuclear exchange between the US and the Soviet Union to break out. And, as the Harvard scholar Graham Allison notes, "What we have learned in the later decades [since the crisis] has done nothing to lengthen those odds."[1] Indeed, history reveals many instances in which a catastrophe was indeed averted, *but just barely*. Our world could have turned out very differently than it did.

Most experts today concur that the threat posed by nuclear weapons is significant: a bomb exploding somewhere in the world is generally understood to be a matter of when rather than if. In his 2004 book *Nuclear Terrorism*, for example, Allison writes, "In my own considered judgment, on the current path, a nuclear terrorist attack on America in the decade ahead is more likely than not."[2] Although this failed to occur—thank goodness—Allison's gloominess is representative of the opinions had by many nuclear experts. For instance, Robert Galluccii, dean of Georgetown University's School of Foreign Service, stated in 2005 that "it is more likely than not that al-Qaeda or one of its affiliates will detonate a nuclear weapon in a US city within the next five to ten years."[3] And the Stanford cryptologist and founder of NuclearRisk.org Martin Hellman puts the probability of a nuclear bomb going off at 1% every year from the present, meaning that "in 10 years the likelihood is almost 10 percent, and in 50

43

years 40 percent if there is no substantial change."[4] An even less optimistic estimate comes from a 2008 report issued by the US Commission on the Prevention of Weapons of Mass Destruction Proliferation and Terrorism, which concludes that "it is more likely than not that a weapon of mass destruction will be used in a terrorist attack somewhere in the world by the end of 2013." More generally, a survey of 85 national security experts in 2005 found that "60 percent of the respondents assessed the odds of a nuclear attack within 10 years at between 10 and 50 percent, with an average of 29.2 percent." And almost 80% of those polled said the attack would come from a terrorist organization.[5]

The risk posed by nuclear terrorism could, in fact, be climbing. As Gary Ackerman and William Potter write in the 2008 book *Global Catastrophic Risks*, it "appears that some prior restraints on terrorist pursuit and use of nuclear weapons . . . may be breaking down."[6] For example, a growing number of terrorists have explicitly stated a desire to use nuclear weapons against their enemies. The late Osama bin Laden referred to the acquisition and detonation of nuclear weapons as his "religious duty," and the *New York Times* reported in 2004 that the CIA overheard al-Qaeda talking about the logistics of creating an "American Hiroshima."[7] While most religious terrorism today is associated with fundamentalist forms of Islam, the threat of a nuclear attack extends far beyond this genre of religiously motivated violence to include "radical right-wing groups," "especially those components espousing extremist Christian beliefs."[8] It also stems from nuclear-armed states, such as the US, Russia, Pakistan, and North Korea, as we'll discuss shortly.

One of the best resources for assessing the threat of nuclear annihilation comes from the *Bulletin of the Atomic Scientists*. This journal was established by a group of physicists who, after having worked on the Manhattan Project (the "big science" enterprise that created the first nuclear bomb), were horrified by the destructive power of the weapons they'd created. In 1947, the *Bulletin* devised a simple way to convey to the public how close humanity is to a catastrophe. Enter the Doomsday Clock. The minute hand of this clock represents our present situation, and midnight a calamity of existential proportions. Over the years, the minute hand has moved toward and away from midnight, back and forth, depending on which way the winds of circumstance were blowing. For example, when the Cold War effectively ended in 1991, the time was set back to 11:43. This is the furthest away from annihilation we've been since the clock was created.

The probability of a global catastrophe has increased in recent years, according to the Doomsday Clock. In 2015, the *Bulletin* moved the minute hand forward from a worrying 5 minutes before midnight to a mere 3, in part because of (as they put it) "global nuclear weapons modernizations, and outsized nuclear weapons arsenals." The official release adds that "world leaders have failed to act with the speed or on the scale required to protect citizens from potential catastrophe. These failures of political leadership endanger every person on Earth."[9] The *only time* since 1945 that the clock has been closer to doom was in 1953, when both the US and the Soviet Union tested thermonuclear weapons. It's worth taking a moment to reflect on this fact: as this book goes to press, the *Bulletin* thinks we're the second-closest to disaster that we've been in more than seventy years.

When Hell Freezes Over

Let's get into a few technical details for a moment. These will provide a helpful context in which to understand the nuclear threat.

To begin, there are two types of nuclear weapons (aside from dirty bombs, which simply involve strapping radioactive material to conventional explosives): *atomic bombs* (or A-bombs) and *hydrogen bombs* (or H-bombs; also called thermonuclear weapons). Both types use as their raw material plutonium (Pu) or uranium (U), the latter of which must be "enriched." This means that the percentage of the U-235 isotope, a variant of uranium containing more neutrons than usual, in the material is *increased*. Whereas the uranium used in nuclear power plants—referred to as "reactor grade"— consists of 3% to 4% U-235, the uranium used in nuclear bombs—referred to as "weapons grade"—is 90% U-235. By comparison, the U-235 found in natural samples of uranium is less than 1%.

The difference between A-bombs and H-bombs concerns how each produces its explosion. A-bombs either compress their material or shoot two clumps together (via the "gun" method) to reach "critical mass," the point at which atoms begin ejecting neutrons. These subatomic particles then collide with other atoms in the vicinity, causing them to eject neutrons, which in turn collide with more atoms, and so on. This "nuclear fission" results in a self-propagating process called a *nuclear chain reaction*. Consequently, huge amounts of energy are released, and this takes the form of an explosion. It's an interesting fact that the bomb detonated above Hiroshima used only about 2% of its uranium: it was, in a sense, more of a dirty bomb than a nuclear bomb.[10]

In contrast, H-bombs employ a combination of nuclear fission *and* nuclear fusion, where nuclear fusion involves different isotopes of hydrogen (H) becoming fused together. Like fission, fusion releases enormous quantities of energy, and in fact H-bombs can be far more powerful than their atomic siblings. The first H-bomb ever built, called "Ivy Mike," was detonated in 1952 by the United States. It was about *700 times as powerful* as the device that fell upon the civilians in Hiroshima.[11] To date, the most powerful bomb ever detonated—the biggest explosion *ever* on Earth— was built by the Soviet Union in 1961. Named the "Tsar Bomba" (or the "emperor bomb"), it produced a blast roughly *5 times as big* as Ivy Mike, although it could have been 10 times more powerful if not for the Soviets reducing its yield to limit the radioactive fallout.

When a nuclear weapon explodes, it generates a ball of *fire* hot enough to vaporize everything in the immediate area. A massive blast wave moving at over 600 miles per hour expands outward in every direction. Depending on how close one is to the weapon's epicenter, this tidal wave of high-density atmosphere can cause the eardrums to rupture, the lungs and abdominal cavity to hemorrhage, and one's joints to be severely strained. Nuclear weapons also release enough thermal energy and ultraviolet light to give people seven miles away first-degree burns. The brilliant flash (actually two quick flashes) it produces can induce temporary blindness lasting up to 40 minutes. In some cases, the retina can be scorched to the point of causing a permanent loss of vision.

The rising hot air after a bomb explodes is capable of sucking radioactive particles tens of thousands of feet into the atmosphere. This produces a horrifying phenomenon called "nuclear fallout." On a global scale, these particles can contaminate food on all continents. On a local scale, people in the blast area can suffer from radiation exposure, which includes damage to the DNA, acute radiation sickness, and cancer. In the case of Hiroshima and Nagasaki, nuclear fallout took the form of a "sticky, dark, dangerously radioactive water" that began to fall from the sky 30 to 40 minutes after the bombs detonated. This substance is known as *black rain*, and it "not only stained skin, clothing, and buildings, but also was ingested by breathing and by consumption of contaminated food or water." Consequently, many people in the region came down with bad cases of radiation sickness.[12]

The blast of a nuclear explosion can also result in a *firestorm*, or a conflagration so huge that it's able to produce and sustain gale-force winds. The bombing of Hiroshima caused a firestorm, which was partly responsible for the death of 70,000 to 80,000 people. If a nuclear weapon were to produce a sufficiently large firestorm, the soot released could become lodged high up in the stratosphere (the layer of atmosphere just above the troposphere, in which we live). These particles would consequently blot out the sun for months or years on end—possibly even *up to a decade*.[13] The result would be a *nuclear winter*, an artificial cooling period during which land temperatures could plunge into the subfreezing range. This would induce catastrophic crop failures, mass famine, widespread starvation, malnutrition, the spread of infectious diseases (in part due to malnutrition), social upheaval, and economic collapse.

We can extrapolate the consequences of a long-term nuclear winter from less severe instances of *volcanic winters*, whose effects, as we'll discuss in chapter 9, are nearly identical to those of a nuclear winter. For example, a volcano that erupted almost two centuries ago produced a cooling period that resulted in agricultural failures, famine, starvation, malnutrition, widespread social unrest, riots, and several major epidemics. If a nuclear weapon were to start a firestorm that spewed enough soot into the atmosphere to block the sun for *a whole decade*, the consequences would be far more apocalyptic. Civilization could be permanently damaged or completely destroyed. At the extreme, since a nuclear winter would affect every area of Earth, a nuclear exchange could bring about the total annihilation of our species.

With God on Our Side

In October 2010, a Japanese artist named Isao Hashimoto posted a video on YouTube, the first of a three-part project to expose "the fear and folly of nuclear weapons."[14] The concept of Hashimoto's project was simple, yet its impact is quite poignant. With an aesthetic reminiscent of early 1990s video games, it condenses the first fifty-three years of the Atomic Age, from 1945 to 1998, into a twelve-minute time-lapse video. A map of the world is displayed in gloomy shades of blue while a metronomic bleep steadily counts up the months, each sound one second apart. Every time a nuclear bomb detonates on an island in the Pacific Ocean, in Africa, or in China, a dot appears along with a beep (different than the metronome sound). At the top and bottom of the screen are the national flags of the countries that have detonated nuclear devices, next to which the total number of bombs are listed.

While the video starts off rather uneventful—one may even be tempted to turn it off, since nothing seems to be happening—by the middle there are so many flashes leaping about the screen and beeps jumping through the speakers that the metronome almost disappears into the background. This continues until the video reaches 1993, at which point it gets *eerily quiet*, with only an occasional disturbance of light and sound. The situation remains fairly calm until the end of the video, at which point the cumulative number of nuclear detonations is shown to be a shocking *2,053*. The only reason a nuclear winter hasn't been induced by such madness is that the bombs were intentionally placed in remote areas where a soot-expelling firestorm couldn't have been started. (Nonetheless, as Greenpeace points out, "every human alive now and over the next tens of thousands of years will carry radioactive elements created by nuclear tests, research and deployment, causing an increase—however small—in their lifetime cancer risk."[15])

The most nuclear weapons our global village has ever housed at once is a mind-blowing 65,000, in 1986. Since this ignominious year, the stockpile has dropped significantly, but what remains is still enough to destroy the planet. As of 2009, the United States had nearly 10,000 nuclear warheads; about 1,500 of these were deployed on either *Minuteman* intercontinental ballistic missiles, submarine-launched ballistic missiles, or in B-52 bombers. As table A shows, Russia has several thousand more nuclear weapons than the US, and more than twice as many intercontinental missiles. There are, in addition to Russia and the US, seven other countries with nuclear arsenals, namely France, the UK, China, Pakistan, India, Israel, and North Korea (which doesn't make an appearance in Hashimoto's video). Although they're estimated to have some 200 nuclear weapons, Israel is one of only a few countries that hasn't signed the Nuclear Non-Proliferation Treaty, along with India, Pakistan, and North Korea.[16] Israel's official position about whether it is the sole nuclear power in the Middle East is equivocal and secretive.

While the Cold War between communism and capitalism subsided with the collapse of the Soviet Union, tensions appear to be on the rise between the US and Russia. In the past few years, the relationship between the most heavily armed nuclear empires has deteriorated. The Russian president Vladimir Putin himself remarked in 2014 that we may be entering a "new Cold War." Meanwhile, North Korea has repeatedly threatened to engage

Table A. Nuclear Weapons, as of 2009[17]

COUNTRY	INTERCONTINENTAL	SHORT-RANGE	BOMBS	SUBMARINES / NONSTRATEGIC	IN RESERVE / AWAITING DISMANTLEMENT	TOTAL IN 2009	TOTAL IN 2000
Russia	1,355	576	856	2,050	8,150	12,987	21,000
United States	550	1,152	500	500	6,700	9,552	10,577
France	–	–	60	240	–	300	350
Israel	–	–	–	–	–	200	0
United Kingdom	–	–	–	192	–	192	185
China	121	–	55	–	–	176	400
Pakistan	–	–	–	–	–	90	0
India	–	–	–	–	–	75	0
North Korea	–	–	–	–	–	2	0

the US in a nuclear exchange. One of the most recent incidents involved a rather mediocre "comedy" starring James Franco called *The Interview*. The movie is about the assassination of North Korea's fearless leader, Kim Jong-un. Declaring it a "movie of terrorism" and "an act of war," North Korea threatened "merciless" retaliation in June 2014 if the movie were released. Although a nuclear strike never occurred, *The Interview* did end up being the target of one of the biggest acts of cybervandalism, or perhaps cyberterrorism, in history.

India and Pakistan also have a long history of conflict that nearly culminated in a 2002 "war that both governments thought might go nuclear."[18] Indeed, many experts recognize South Asia as the most probable region for a nuclear war to occur. The result of a nuclear exchange between these countries would be devastating for the locals, since the population density of the region is far greater than the density of, for example, North America and Europe.[19] Consequently, more people would die from weapons of the same explosive yield. A nuclear war between India and Pakistan could also potentially induce a nuclear winter, thereby affecting people in every corner of the world. According to one study, a regional "nuclear exchange involving 100 Hiroshima-size bombs (15 kilotons) on cities in the subtropics" could effectively "lower temperatures regionally and globally for several years, open up new holes in the ozone layer protecting the Earth from harmful radiation, reduce global precipitation by about 10% and trigger massive crop failures."[20]

But Pakistan is worrisome for other reasons as well. As table A shows, they have some 90 nuclear weapons, and in the past decade their government has become increasingly unstable. For example, while tripling its nuclear arsenal, Pakistan simultaneously failed to prevent the leadership of al-Qaeda from making its northern territories a home base.[21] By 2015, many fighters in the region had joined the Islamic State, after this group "created a 10-man 'strategic planning wing' with a master plan on how to wage war against the Pakistani military."[22] These developments are extremely worrisome because Pakistan has a track record of failing to keep its nuclear program secure. The most notable instance of a major security failure came from a now-infamous physicist named Abdul Qadeer Khan, who surreptitiously sold designs for nuclear bombs and related technology to North Korea, Iran, and Libya. Khan was arrested in 2004, but the conditions that enabled his malfeasance have not been properly addressed or fixed. (Incidentally, Khan was released from house arrest in 2009 and is now a free man.) It is therefore entirely possible that "opportunities for mini-Khans to proliferate nuclear technology" to groups like the Islamic State arise in the future.[23] As the Commission on Prevention of Weapons of Mass Destruction Proliferation and Terrorism put it, "Were one to map terrorism and weapons of mass destruction today, all roads would intersect in Pakistan."[24]

Radical Islamists aren't the only terrorists with nuclear ambitions. There's also a growing threat from right-wing groups, many of which are associated with extremist Christian ideologies. For example, the Christian Identity movement, mentioned in chapter 1, is an apocalyptic group "whose faith entails a deep belief in the need to cleanse and purify the world via violent upheaval to eliminate nonbelievers."[25] As Charles Ferguson and William Potter write in their book *The Four Faces of Nuclear Terrorism*, "These types of groups, driven by an urgency and religious passion, often have characteristics . . . that make them of great concern as potential nuclear terrorists."[26] Other terrorist groups with Christian leanings include the Ku Klux Klan, the Order, Defensive Action, the Freemen Community, and the Aryan Nations. In fact, the man behind the Oklahoma City bombing, Timothy McVeigh, was likely influenced by Christian Identity doctrines, which he was exposed to "through the militia culture with which he was associated."[27] Before 9/11, the Oklahoma City bombing was the most catastrophic terrorist attack in the United States.

According to the former director general of the International Atomic Energy Agency (IAEA), an intergovernmental organization dedicated to promoting "atoms for peace," "nuclear terrorism is the most serious danger the world is facing." This sentiment—echoing Allison and his colleagues above—may be true, but it's crucial to note that the threat of a nuclear apocalypse doesn't just stem from individuals with AK-47s in the deserts of modern Mesopotamia. There are many people we wouldn't classify as terrorists who are nontrivially increasing the chance of nuclear war. For example, in chapter 13, we'll examine how the "Armageddon lobby" in the US has been pushing for a confrontation with Iran. One of the most influential leaders of this group has, in fact, repeatedly and explicitly called for a nuclear first strike against the Iranian state. In this way, these "dispensationalists" constitute a significant threat to world peace, just like the jihadists in Iraq and Syria.

Close Calls and Near Misses

So far, we've focused exclusively on the threat posed by terror. But there's an error side to the dual-use coin as well. That is to say, *even if* our species behaves itself and manages to avoid self-immolation, we could still end up starving under the dark skies of a nuclear winter. As it happens, history reveals numerous instances of nuclear almost-disasters resulting from mistakes, accidents, negligence, and technical failures. For example, on January 24, 1961, a B-52 Stratofortress was circling over the East Coast with two hydrogen bombs in its belly. It was part of the Chrome Dome mission, which aimed to keep nuclear weapons in the air at all times, in case the Soviet Union attacked without warning. Midflight, the bomber began to leak large amounts of kerosene from its right wing. Within a minute and a half, one of its fuel tanks emptied, after which its main tank began to lose fuel. The pilots directed the plane toward a US Air Force base in Wayne County, North Carolina.

As it approached the runway with its landing gear down, between 2,000 and 10,000 feet in the air, the aircraft exploded.[28] Both hydrogen bombs onboard were released into the air: one plowed 20 feet into the ground at a speed of 700 mph.[29] The other had its parachute open and, despite initial reports from the Air Force to the contrary, went through *six of the seven steps* of the firing sequence leading up to detonation. In the end, neither bomb exploded, although "a portion of one weapon, containing uranium, could not be recovered despite excavation in the waterlogged farmland to a depth of 50 feet."[30] If one of the bombs had gone off, it could have led the US to mistakenly believe it was being attacked, thus precipitating a retaliatory strike against the Soviet Union. This is the infamous Goldsboro Crash.[31]

A more recent incident occurred in 2007, when a B-52 flew from North Dakota to an air base in Louisiana. Nothing was notable about this flight except that the crew were completely unaware that the plane contained six nuclear-armed missiles. The pilots should have been informed by Air Force weapons officers, but the officers repeatedly failed—five times, in fact—to examine the bomber's cargo. Once the plane landed in Louisiana, it sat on the runway for twenty-four hours without proper security. A total of 70 airmen were punished because of this blunder, one that could have resulted in a nuclear disaster of some sort (if the plane had crashed) or, potentially, a missile being stolen.[32]

Many more examples like these could be adduced. For example, two nuclear torpedoes likely came to rest somewhere on the bottom of the Atlantic Ocean after an American submarine carrying them mysteriously vanished. What probably happened, according to a 1969 Naval Court report first made public in 1993, is that during a practice exercise one of the nonnuclear torpedoes became activated. In response, the crew released it from the submarine. Unfortunately, the torpedo's homing device became activated, thereby causing it to loop around and hit the submarine from which it came.[33] And in 1983, a man named Stanislav Petrov quite possibly saved the world by correctly identifying two early-warning-system reports of nuclear missiles heading from the US to the Soviet Union as erroneous. In 2014, a feature film about this incident premiered at the Woodstock Film Festival, aptly titled *The Man Who Saved the World*.

As for the three missile-launch facilities containing LGM-30 Minuteman nuclear missiles in the US, none are in good shape. According to a CBS report, the missile silo in Wyoming contains a broken door leading to the capsule. With a danger sign hanging on it, it must be propped open with a crowbar. The computers at this facility responsible for receiving launch orders from the US president are also outdated, relying on floppy disks from the 1980s and 1990s. Meanwhile, at the Malmstrom Air Force Base in Montana, a launch officer apparently texted answers to a monthly proficiency test to other officers. As the *New York Times* reports, "nearly half of the missile launch crew's members either were cheating on monthly proficiency tests or knew about the cheating."[34] In the aftermath of this scandal, which involved "100 or so officers," 34 officers lost their certification and 9 were fired.[35] And in North Dakota, 17 Air Force officers responsible for launching nuclear missiles had their launch keys revoked after performing poorly during an inspection. As ABC reports, "one of the officers was investigated for potentially compromising nuclear launch codes, a decision on disciplinary action is pending."[36]

So, the threat posed by *nuclear terrorism* is both nontrivial and on-going. As long as countries maintain large arsenals of nuclear weapons, the grim possibility that one will go off by mistake will continue to haunt us. At worst, an error of some sort could trigger a nuclear exchange between one or more states. As I write this (but hopefully not as you read it), the US and Russia still have some 2,000 nuclear missiles aimed at each other.[37] (To be specific, these missiles aren't actually aimed at each other's cities, but they could be in a matter of 15 minutes or so.[38]) A couple of blasts could not only kill millions of people but initiate winter-inducing firestorms capable of jeopardizing the survival of intelligent life on Earth. During the 2009 United Nations Security Council Summit on Nuclear Non-Proliferation and Nuclear Disarmament, Barack Obama noted, correctly, that a *single* nuclear device exploding anywhere on Earth "would badly destabilize our security, our economies, and our very way of life."[39] In a phrase: it may be that the biggest shadows in the nuclear valley of death extend from the peaks of error, not just those of terror.

Duck and Cover

The Atomic Age is not irreversible: there are concrete steps we can take to eliminate the threat of nuclear weapons. One is to implement treaties that would limit, or possibly even eliminate, nuclear arsenals. Yet there's general agreement among "officials, diplomats, think-tankers, and scholars" that the Nuclear Non-Proliferation Treaty "is in bad shape, in danger, in crisis, or eroding."[40] In 2005, for example, the Bush-appointed

US ambassador to the United Nations, John Bolton, declared that thirteen steps toward nuclear disarmament previously agreed upon in 2000 were impractical. Bolton then forbade any use of the word "disarmament" from the Review Conference's "outcome document."[41] Incidentally, two years earlier, the Bush administration caused some controversy by pushing for the development of new, low-yield weapons called "nuclear bunker busters," which can penetrate deep into the ground before detonating. The concern among experts was that such weapons would have blurred the lines between nuclear and conventional warfare, thereby opening the door to future conflicts involving larger nuclear devices. Fortunately, the plans were retracted in 2005.

History shows that agreements like the Nuclear Non-Proliferation Treaty can be effective to some extent, at least between nation-states. The Megatons to Megawatts Program, for example, was an arrangement that started in 1993 whereby Russia converted highly enriched uranium from "the cores of nuclear warheads originally designed to destroy American cities" into low-enriched uranium for use in US nuclear power plants. It's a largely unknown fact that about half of all the electricity produced by nuclear power between 2000 and 2010 was fueled by repurposed uranium once intended to obliterate the West.[42] With respect to terrorism, robust countermeasures could effectively mitigate the risks posed by malevolent groups. The subtitle to Allison's aforementioned book is, in fact, *The Ultimate Preventable Catastrophe*. But neutralizing terrorism will require a significant, concerted effort among world governments. Between 1993 and 2013, a total of 664 incidents involving "theft or loss of nuclear or other radioactive material" were reported to the IAEA.[43] This is cause for serious concern, and it must be corrected.

While some analysts would argue that a world with nuclear weapons is actually *safer* than one without,[44] it appears trivially true that as long as such weapons exist, nuclear error and terror will pose a major threat to our future.

Notes

1. See Graham Allison, "The Cuban Missile Crisis at 50," *Foreign Affairs* (July/August 2012), https://www.foreignaffairs.com/articles/cuba/2012-07-01/cuban-missile-crisis-50.

2.   See Graham Allison, *Nuclear Terrorism: The Ultimate Preventable Catastrophe* (Owl Books, 2004), p. 15.

3.   See Robert Gallucci, "Averting Nuclear Catastrophe," *Harvard International Review*, January 6, 2005, http://hir.harvard.edu/archives/1303.

4.   See Dan Farber, "Nuclear Attack a Ticking Time Bomb, Experts Warn," *CBS News*, May 4, 2010, http://www.cbsnews.com/news/nuclear-attack-a-ticking-time-bomb-experts-warn/.

5.   Again, see Dan Farber, "Nuclear Attack a Ticking Time Bomb, Experts Warn," *CBS News*, May 4, 2010, http://www.cbsnews.com/news/nuclear-attack-a-ticking-time-bomb-experts-warn/.

6.   See Gary Ackerman and William Potter, "Catastrophic Nuclear Terrorism: A Preventable Peril," in *Global Catastrophic Risks*, ed. Nick Bostrom and Milan Ćirković (Oxford University Press, 2008), p. 417.

7.   See Nicholas Kristof, "An American Hiroshima," *New York Times*, August 11, 2004, http://www.nytimes.com/2004/08/11/opinion/an-american-hiroshima.html.

8.   See Gary Ackerman and William Potter, "Catastrophic Nuclear Terrorism: A Preventable Peril," in *Global Catastrophic Risks*, ed. Nick Bostrom and Milan Ćirković (Oxford University Press, 2008), p. 418.

9.   See "Press Release: It Is Now 3 Minutes to Midnight," *Bulletin of the Atomic Scientists*, January 22, 2015, http://thebulletin.org/press-release/press-release-it-now-3-minutes-midnight7950.

10.   See "Nuclear 'Command and Control': A History of False Alarms and Near Catastrophes," *NPR*, August 11, 2014, http://www.npr.org/2014/08/11/339131421/nuclear-command-and-control-a-history-of-false-alarms-and-near-catastrophes.

11.   See "1 November 1952—Ivy Mike," CTBTO Preparatory Commission, https://www.ctbto.org/specials/testing-times/1-november-1952-ivy-mike.

12.   See "Energy and Radioactivity," Atomic Bomb Museum, http://atomicbombmuseum.org/3_radioactivity.shtml.

13.   See Joseph Cirincione, "The Continuing Threat of Nuclear War," in *Global Catastrophic Risks*, ed. Nick Bostrom and Milan Ćirković (Oxford University Press, 2008), pp. 391–392.

14.   See "A Time-Lapse Map of Every Nuclear Explosion Since 1945," YouTube video, uploaded October 24, 2010, https://www.youtube.com/watch?v=LLCF7vPanrY.

15. See "What's the Damage?" GreenPeace, April 26, 2006, http://www.greenpeace.org/international/en/campaigns/peace/abolish-nuclear-weapons/the-damage/.

16. Note that North Korea has joined and then withdrawn from the Nuclear Non-Proliferation Treaty *twice*. See "Nuclear Weapons: Who Has What at a Glance" and "Chronology of US-North Korea Nuclear and Missile Diplomacy," both published by the Arms Control Association.

17. This diagram is taken from Julian Borger, "Nuclear Weapons: How Many Are There in 2009 and Who Has Them?" *Guardian*, September 25, 2009, http://www.theguardian.com/news/datablog/2009/sep/06/nuclear-weapons-world-us-north-korea-russia-iran#data

18. See Graham Allison, "Nuclear Disorder: Surveying Atomic Threats," *Foreign Affairs* (February 2010), http://www.foreignaffairs.com/articles/65732/graham-allison/nuclear-disorder.

19. See Joseph Cirincione, "The Continuing Threat of Nuclear War," in *Global Catastrophic Risks*, ed. Nick Bostrom and Milan Ćirković (Oxford University Press, 2008), p. 390.

20. See Joseph Cirincione, "The Continuing Threat of Nuclear War," in *Global Catastrophic Risks*, ed. Nick Bostrom and Milan Ćirković (Oxford University Press, 2008), p. 392.

21. See Graham Allison, "Nuclear Disorder: Surveying Atomic Threats," *Foreign Affairs* (February 2010), http://www.foreignaffairs.com/articles/65732/graham-allison/nuclear-disorder.

22. See Mujeeb Ahmed, "ISIS Has Master Plan for Pakistan, Secret Memo Warns," *NBC News*, November 10, 2014, http://www.nbcnews.com/storyline/isis-terror/isis-has-master-plan-pakistan-secret-memo-warns-n244961.

23. See Graham Allison, "Nuclear Disorder: Surveying Atomic Threats," *Foreign Affairs* (February 2010), http://www.foreignaffairs.com/articles/65732/graham-allison/nuclear-disorder.

24. See Ewen MacAskill, "US Report Predicts Nuclear or Biological Attack by 2013," *Guardian*, December 3, 2008, http://www.theguardian.com/world/2008/dec/03/terrorism-nuclear-biological-obama-white-house.

25. See Charles Ferguson and William Potter, *The Four Faces of Nuclear Terrorism* (Routledge, 2005), p. 18.

26. Again, see Charles Ferguson and William Potter, *The Four Faces of Nuclear Terrorism* (Routledge, 2005), p. 18.

27.   See Mark Juergensmeyer, *Terror in the Mind of God: The Global Rise of Religious Violence* (University of California Press, 2003), p. 31.

28.   See "The Full Story," ibiblio.org, http://www.ibiblio.org/bomb/initial.html.

29.   See Rudolph Herzog, *A Short History of Nuclear Folly* (Melville House, 2013), p. 200.

30.   See "The Full Story," ibiblio.org, http://www.ibiblio.org/bomb/initial.html.

31.   This paragraph relied heavily on Rudolph Herzog, *A Short History of Nuclear Folly* (Melville House, 2013), pp. 199–202.

32.   See "Air Force Punishes 70 in Accidental Nuclear-Weapons Flight," *Arkansas Online*, October 19, 2007, http://www.arkansasonline.com/news/2007/oct/19/air-force-punishes-70-accidental-nuclear-weapons-f/.

33.   See Peter Limburg, *Deep-Sea Detectives: Maritime Mysteries and Forensic Science* (ASJA Press, 2005), pp. 194–195.

34.   See Helene Cooper, "Air Force Fires 9 Officers in Scandal Over Cheating on Proficiency Tests," *New York Times*, March 27, 2014, http://www.nytimes.com/2014/03/28/us/air-force-fires-9-officers-accused-in-cheating-scandal.html?_r=1.

35.   Again, see Helene Cooper, "Air Force Fires 9 Officers in Scandal Over Cheating on Proficiency Tests," *New York Times*, March 27, 2014, http://www.nytimes.com/2014/03/28/us/air-force-fires-9-officers-accused-in-cheating-scandal.html?_r=1. See also, Greg Botelho, "9 Air Force Commanders Fired from Jobs over Nuclear Missile Test Cheating," *CNN*, March 27, 2014, http://edition.cnn.com/2014/03/27/us/air-force-cheating-investigation/.

36.   See Luis Martinez, "17 Air Force Officers Have Nuke Launch Keys Revoked," *ABC News*, May 8, 2013, http://abcnews.go.com/blogs/politics/2013/05/17-air-force-officers-have-nuke-launch-keys-revoked/. For a wonderfully comical take on all this, see "Nuclear Weapons," *Last Week Tonight with John Oliver*, July 27, 2014, https://www.youtube.com/watch?v=1Y1ya-yF35g.

37.   See Tony Magliano, "Nuclear Weapons Still Threaten the World," *Catholic Online*, August 1, 2014, http://www.catholic.org/news/hf/faith/story.php?id=56371.

38.   Thanks to Tony Barrett and Joseph Siracusa for clarifying this point for me.

39.   See "Remarks by the President At the UN Security Council Summit," White House, September 24, 2009, https://www.whitehouse.gov/the_press_office/Remarks-By-The-President-At-the-UN-Security-Council-Summit-On-Nuclear-Non-Proliferation-And-Nuclear-Disarmament.

40.   See Liviu Horovitz, "Beyond Pessimism: Why the Treaty on the Non-Proliferation of Nuclear Weapons Will Not Collapse, *Journal of Strategic Studies* 38, no. 1–2 (2015).

41.   See Graham Allison, "Nuclear Disorder: Surveying Atomic Threats," *Foreign Affairs* (February 2010), http://www.foreignaffairs.com/articles/65732/graham-allison/nuclear-disorder.

42.   Here I'm paraphrasing Graham Allison, "Nuclear Disorder: Surveying Atomic Threats," *Foreign Affairs* (February 2010), http://www.foreignaffairs.com/articles/65732/graham-allison/nuclear-disorder.

43.   See "Indicent and Trafficking Database (ITDB)," International Atomic Energy Agency, http://www-ns.iaea.org/security/itdb.asp.

44.   See Joseph Siracusa, *Nuclear Weapons: A Very Short Introduction* (Oxford University Press, 2008), p. 108.

3

Bugs

• • • • •

Pandemonium

What caused more human suffering, World War I or the Spanish flu of 1918? The former lasted four years, from 1914 to 1918, and the latter only two years, from 1918 to 1920. When asked, most people answer that—of course!—World War I was the greater of the two evils, if measured in the currency of human life. Indeed, many people haven't even heard of the Spanish flu pandemic. But the surprising fact is that in only a couple of years, the flu pandemic cast up to *63 million more* people into the muddy earth than all the bayonets, bullets, and bombs of World War I; it would take roughly *three* First World Wars to equal the death caused by the 1918 influenza pandemic. Such statistics are not anomalous when one compares outbreaks of infection to outbreaks of war. For example, the Black Death, an epidemic of the plague that devastated fourteenth-century Europe, took roughly as many people as the Thirty Years' War, the Russian Civil War, the Mongol conquests, the Crusades, World War I, and World War II *combined*.

The astonishing reality is that pathogenic microorganisms—or *bugs*— have inflicted more misery, suffering, and death upon our species than we have upon ourselves. In the process, they've shaped the course of human history in significant ways. The Plague of Cyprian, for example, was an outbreak of (what is now believed to have been) smallpox in the third century. It was said to have killed 5,000 people in a single day in Rome.*

---

* Incidentally, this outbreak of death fit Jesus' prophecy, and, as a result, many Christians were overcome with "unimaginable joy" about the misery around them. As Cyprian gleefully wrote, "The kingdom of God, most beloved brothers, has begun to be imminent."

What's perhaps most interesting about this episode of anguish is that it may have been responsible, in a nontrivial way, for the rise of Christianity, since adherents of the young religion decided to "martyr" themselves rather than die from the disease, which appeared a likely outcome. This made Christianity look to nonadherents like it was worth dying for, causing many to convert. Thus, counterfactually speaking, if the Plague of Cyprian had never occurred, Christianity might not have become the religious superpower it now is.[1]

Another example of bugs shaping history is the decimation of the Native American population by the Europeans in the centuries after Christopher Columbus sailed to the Bahamas. One of the hallmarks of large societies is, in fact, their significant "burden of disease." That is to say, "civilized" peoples tend to be far *sicker* than their "primitive" counterparts, and it's this asymmetry of illness that's helped empires conquer and destroy small groups with relative ease. Perhaps if we were ever visited by an extraterrestrial civilization, we'd die purely as a result of disease rather than malicious intent on their part.

Pathogenic germs are a ghost that haunts our past, but they're also a specter lurking in the future. While pre-twentieth-century humans didn't have the vast resources of modern medicine to prevent and mitigate diseases of infection (e.g., vaccines and antibiotics), they also weren't living in high-density urban centers and dependent upon highly convoluted networks of international travel and global trade, all of which can facilitate the propagation of pathogens. The 2014 outbreak of Ebola, for example, saw the virus spread from Liberia to the United States quite literally at the speed of a jetliner. In the modern world, 3 million people board airplanes *every day* to reach destinations hundreds or thousands of miles away. Meanwhile, food is shipped around the world and back again, as a growing number of countries are becoming significant importers and exporters of food items.[2] And huge numbers of people live in close quarters—a factor that was largely responsible for the Spanish flu's spread through a whopping 33% of the global population.

The possibility thus remains that nature—the original genetic engineer and a constant tinkerer with life's building blocks—could create a new pathogen as deadly as the Spanish flu, the Black Death, or Ebola. Fortunately, the most lethal pathogens tend not to win the evolutionary game, since killing the host too quickly can prevent the germ from effectively spreading to other organisms, which is necessary for its continued survival in the wild. The host becomes a coffin for the bugs inside it—a literal *dead end*. There is, in other words, a natural check on the virulence of pathogens that evolve in nature.

A Dual Etiology: Turning Biology against Itself

For the vast majority of our species' history, outbreaks of disease arose entirely from nature: this was their one and only etiology. But the situation is different today. Unlike the risks posed by nuclear weapons (which arise entirely from *human activity*) and supervolcanoes (which arise entirely from *nature*), an outbreak of infectious disease can be both natural and anthropogenic in origin. It's the only risk discussed in this book with a *dual etiology*.[3] The fact that a pandemic can have two distinct causes is partly what makes it one of the biggest threats facing humanity this century.

The addition of humans as an etiology of infections is a fairly recent development.[4] Today, it's associated with two interrelated fields of emerging research, namely *biotechnology* and *synthetic biology*. The former includes the field of genetic engineering, in which scientists directly modify the structure, function, and behavior of organisms by adding new genes to their genomes and "knocking out" existing genes so that they no longer function properly. RNA interference is another example of biotechnology. It involves using RNA molecules to inhibit the expression of specific genes, a method that could be used to treat diseases like HIV and cancer. In 2006, the two scientists who pioneered RNA interference received the Nobel Prize in Physiology or Medicine, which indicates just how important this phenomenon is.

Whereas biotechnology aims to manipulate and recombine already existing structures into novel products, synthetic biology strives to design entirely new forms from the "ground up."[5] In 2010, a team of scientists led by the geneticist Craig Venter successfully created the first instance of *synthetic life*. Venter's team did this by copying the genome of a naturally occurring bacterium into a computer, making a few changes to it (such as adding "watermarks"), synthesizing the new DNA sequence in the laboratory from scratch, and then injecting it into a recipient cell whose genetic material had been previously removed.[6] The resulting bacterium—nicknamed "Synthia"—replicated over a billion times. This was an extraordinary accomplishment, and it's paved the way for the team's next goal: to figure out the *minimum* number of genes required for a cell to remain alive. Once such a bacterium is created, additional genes can be added one by one to make the cell perform specific tasks of our choosing, such as "more efficiently and economically produc[ing] vaccines, pharmaceuticals, biofuels, food and other products."[7] Synthetic biology is all about turning cells into tiny programmable computers that can rearrange the world in desirable ways.

Without a doubt, these two burgeoning fields could improve the human condition immensely. The knowledge and products produced by them promise to revolutionize medicine by offering new cures for disease, provide clean energy and energy security, and even reverse the damage we've already done to the environment (for example, by creating bacteria that can absorb carbon dioxide from the atmosphere). This is why, as of 2011, the synthetic biology industry was valued at $1.6 billion, and recent studies suggest that by 2016 it will have ballooned to $10.8 billion. Synthetic biology is "a future major industry," one that governments around the world are investing heavily in, and for good reason.[8]

But as with nuclear technologies, *the silver cloud has a dark lining*: great peril will attend great benefits. As a 2006 paper by the Institute of Medicine and National Research Council states, "it is an unfortunate reality that almost all advances in life sciences technology pose potential 'dual-use' risks."[9] That is, if used in morally responsible ways, such technologies could transform the world into a far more livable place for our children. But the very same entities could be twisted by nefarious individuals to bring about unprecedented levels of suffering—perhaps even to initiate a red-dot or black-dot catastrophe. The resulting "dual-use dilemma," as the National Research Council's "Fink Report" states, poses a serious threat to our future: "the same technologies can be used legitimately for human betterment and misused for bioterrorism."[10] There appears to be no getting around this fact, since the good and bad are a package deal.

As noted in chapter 1, dual usability wouldn't be a cause for concern if the capacity for the relevant technologies to rearrange the world were limited within safe boundaries. In the case of biotechnology and synthetic biology, though, both are growing more *powerful* along what appears to be an exponential trajectory. Our ability to manipulate the microscopic world by toying with nucleic acids today is profoundly greater than it was just a decade ago (indeed, the structure of DNA wasn't even known until 1953). What's worrisome is that the "explosive growth phase" of these fields, to quote an unclassified CIA document, could lead to the creation of bugs that are far more devastating than anything previously cooked up in the kitchen of nature.[11] As mentioned earlier, natural selection imposes upper bounds on the virulence of germs. But a designer pathogen is free of such constraints: it need not contain any *selfish genes* whose primary concern is their own survival in a world of scarce resources. Instead, it could be designed specifically *to kill*.

A more theoretical possibility is that synthetic biological techniques are employed to build a completely alien pathogen, an agent of infection unlike anything the biosphere has ever before seen. As the aforementioned CIA document notes, "growing understanding of the complex biochemical pathways that underlie life processes has the potential to enable a class of new, more virulent biological agents engineered to attack distinct biochemical pathways and elicit specific effects. . . . The same science that may cure some of our worst diseases could also be used to create the world's most frightening weapons."

Making matters worse, synthetic biology and biotechnology are simultaneously becoming more *accessible* to nonexperts and experts alike. Recall that there are four distinct axes along which accessibility can be measured: intelligence, knowledge, skills, and equipment. It appears, with very little ambiguity, that the life sciences are growing along *all four axes*.

For example, as Jonathan Tucker notes in *The New Atlantis*, "de-skilling has already occurred in several genetic-engineering techniques that have been around for more than twenty years." Examples include "gene cloning (copying foreign genes in bacteria), transfection (introducing foreign genetic material into a cell), ligation (stitching fragments of DNA together), and the polymerase chain reaction, or PCR (which makes it possible to copy any particular DNA sequence several million-fold)." He adds that "a few standard genetic-engineering techniques have been de-skilled to the point that they are now accessible to undergraduates and even advanced high school students, and could therefore be appropriated fairly easily by terrorist groups."[12]

Many commentators have noted that if Ted Kaczynski, the Unabomber, had earned a master's in microbiology rather than a PhD in mathematics (trained at Harvard no less), he could have wreaked far greater havoc on society than he did. But one might not even need a master's degree these days. The ambitious autodidact can find a wealth of information online, including lectures, textbooks, videos, and whole genomes of pathogens like polio and smallpox. If you, dear reader, have an extra minute, go ahead and browse the genetic sequence of the Ebola virus—with its ghastly 90%

fatality rate in the absence of treatment—at this link: http://www.ncbi. nlm.nih.gov/bioproject/PRJNA257197. Anyone can access this, from the professional biologist to an Islamic State sympathizer.

The creation of bugs with unprecedented virulence isn't a merely *speculative* possibility. We know it can happen. A famous example occurred in 2001, when a group of Australian scientists attempting to create a mouse contraceptive vaccine accidentally produced a variant of the mousepox virus—which is very similar to the smallpox virus—that was extraordinarily lethal, killing even those mice who'd been vaccinated against it. This suggested to many observers that the smallpox virus could, similarly, be modified to be even more deadly than it already is. As a landmark document published in early 2015 by the Global Challenges Foundation notes, it's conceivable that "all the features of an extremely devastating disease [that] already exist in nature" could be combined, including the incurability of Ebola, the fatality of rabies, the infectiousness of the common cold, and the long incubation periods of HIV. The section concludes that "if a pathogen were to emerge that somehow combined these features . . . its death toll would be extreme."[13]

A year after the mousepox debacle, scientists at Stony Brook University managed to synthesize "a live polio virus from chemicals and publicly available genetic information." Polio is a childhood disease that left huge numbers of young people paralyzed in the late nineteenth and twentieth centuries. It underwent a significant decline after a vaccine was developed in the 1950s. Without relying on any hard-to-get supplies, this team succeeded in building "the virus using its genome sequence, which is available on the Internet, as their blueprint and genetic material from one of the main companies that sell made-to-order DNA." As it happens, the project was funded by the Pentagon, and the aim was "to send a warning that terrorists might be able to make biological weapons without obtaining a natural virus." As the project's leader ominously reports, "You no longer need the real thing in order to make the virus and propagate it."[14]

Since 2002, a startling variety of some of the most dangerous pathogens on Earth have been resurrected from the dead by scientists hoping to develop more effective treatments for disease. Examples include "a SARS-like virus and the formerly extinct strain of influenza virus" that was responsible for the 1918 pandemic of Spanish flu.[15] Before the Spanish flu virus was synthesized, the only remaining samples were preserved as "small DNA fragments in victims buried in Alaskan permafrost, or in tissue specimen of the United States Armed Forces Pathology Institute."[16] It's thus entirely conceivable that the bioterrorist of the future with minimal skills and subgenius intellectual abilities could acquire samples of the most lethal microorganisms on the planet, possibly altering them to enhance virulence, increase contagiousness, and delay the onset of symptoms to maximize its propagation through the human population—or even the population of livestock, to initiate food shortages.

In 2007, the physicist Freeman Dyson described a possible future in which de-skilling has occurred to such an extent that even children can engage in synthetic biology experiments. Recent years have, in fact, seen an emerging movement among nonexperts that's paving the way for Dyson's vision. This movement is called *biohacking*, and it involves amateur hobbyists cobbling together homebuilt laboratories to conduct experiments with microorganisms. While most biohackers are motivated by curiosity, a desire for knowledge, or the hope of creating a cure for disease, others may attempt to create pathogens to show off their "technical prowess" or to perpetrate a bioterrorist attack. The amazing fact is that today, one needs only a few thousand dollars to set up a fully functional laboratory at home—and the cost is dropping precipitously. Many of the processes in these labs are automated (as noted earlier), meaning that one needs only a finger capable of pushing a button to operate them. And virtually any question one might have about how to perform some experiment can be found with a simple Google search.

Furthermore, as the Stony Brook experiment showed, the raw materials needed to conduct avocational research can be simply ordered from commercial providers. In 2006, journalists working for the *Guardian* managed to order "part of [the] smallpox genome through mail order."[17] And while weapons like nuclear bombs require rare materials, namely uranium and plutonium, the raw materials of the microbiology lab are self-replicating and therefore in abundant supply once acquired. A reasonably competent hobbyist with an Internet connection could potentially download the Ebola virus' genetic sequence, order segments from online suppliers, and assemble the parts in an attempt to invent a vaccine. The biohacking terrorist, meanwhile, could attempt to modify the properties of the virus to make it even more dangerous than it is.[18]

What's most disconcerting about the accessibility trend is that lone wolves are extremely hard to detect. They can be invisible to even the most watchful eyes. And there could be a large number of them: statistically speaking, just as there are far more terrorist groups than rogue states, so too are there far more deranged individuals than terrorist groups. In other words, the Ted Kaczynskis of the world far outnumber the Islamic States, and the Islamic States outnumber the North Koreas. The growth of biotechnology and synthetic biology along the four axes of accessibility will therefore make it increasingly likely that an engineered pandemic occurs at some point in the future. This elevating threat level could have the effect of pushing governments to adopt increasingly draconian policies of intrusive regulation, in the name of "security for all." Consequently, individual privacy could be severely compromised by surveillance apparatuses designed to monitor every move one makes, leading to the emergence of oppressive totalitarian regimes, itself a possible existential catastrophe of the endurable variety.[19]

Yet without such intrusive surveillance, humanity will become more vulnerable to unprecedented attacks by people or small groups (e.g., "sleeper cells") that we never even knew were threats. If such actors were to release a designer pathogen into the London Underground or spread it across a high-density slum (e.g., those around Mumbai, Mexico City, or Nairobi), huge numbers of people could perish. The catastrophe could have global repercussions, including social upheaval, economic disruptions, agricultural failures, and significant "damage to the world's trade network."[20] At the extreme, the pathogen could spread to nearly the entire human population, thereby threatening our species with extinction.

While the international community's response to the recent eruptions of Ebola, SARS, and Swine Flu demonstrate that our strategies for guarding against nasty bugs can be effective, there's no telling how well humanity could contain, for example, a variant of the smallpox virus created for the sole purpose of causing death. A federal investigation in September 2014 found that the United States Department of Homeland Security is alarmingly "ill-prepared" for an emergency of infectious disease. In fact, it discovered huge stockpiles of hand sanitizer and antiviral drugs that had expired, as well as unknown quantities of lost and potentially unusable protective gear. The United States is one of the most developed regions of the world, yet if an outbreak were to occur here, the report states, it could "cause significant illnesses and fatalities, and substantially disrupt our economic and social stability."[21] This is an ominous sign.

As former UN Secretary-General Kofi Annan stated in a 2006 talk at Princeton University, the possibility of an engineered pandemic—or some other version of bioterrorism—constitutes "the most important, under-addressed threat relating to terrorism—one which acutely requires new thinking."[22] This sentiment has been echoed by *many* experts: we are, as a National Research Council committee on biosecurity puts it, approaching yet another new "threat horizon" in the wilderness of existential risk.

Menacing Mistakes

As with all dual-use artifacts, the "harmful" category of use divides into error and terror. So far, we've been fixated entirely on the threat of terror. Let's now turn to the possibility that an accident, blunder, or mistake leads to a large-scale disaster involving bugs.

We mentioned biohackers in the previous section, many of whom are driven to "hack into living systems" by sincere curiosity, or the human impulse for knowledge. The problem is that the more people there are handling microorganisms—some of which may be pathogenic—the more chance there is, statistically, for *someone* to screw up. This possibility of error could pose a significant risk to public health. But *even with* proper governmental oversight, accidents involving highly virulent germs in highly regulated labs happen on a shockingly frequent basis. For example, a recently released government report catalogues over *1,100 laboratory blunders* involving hazardous biomaterials between 2008 and 2012. In one incident, "two animals were accidentally infected with hog cholera, a virus that hasn't been found in the US since 1978." In another, a "cow residing on a nearby farm became infected with brucellosis, a virus that can be passed to humans through dairy products."[23]

In 2014, the *New York Times* reported that during the first two weeks of June, up to 75 scientists at the Centers for Disease Control (CDC) "may have been exposed to live anthrax bacteria after potentially infectious samples were sent to laboratories unequipped to handle them."[24] Several employees were treated with antibiotics and, luckily, no one became ill. The following month, another incident was made public: this time "a CDC lab accidentally contaminated a relatively benign flu sample with a dangerous H5N1 bird flu strain that has killed 386 people since 2003."[25] Around the same time as these debacles, a few employees working on the Food and Drug Administration's Bethesda, Maryland, campus stumbled upon some misplaced samples of the smallpox virus in an unused storage room. Incredibly, the specimens, contained in two vials dating back to 1954, were still *viable*, meaning that they were actively growing and capable of infecting anyone exposed to them.[26] More than 300 vials of pathogens were discovered in the same room, "including the virus behind the tropical disease dengue and the bacteria that can cause spotted fever," as well as influenza and Q fever.[27] The CDC is, as its director puts it, "the reference laboratory to the world." Yet despite the caliber of scientists working there and the regulatory oversight, it has repeatedly failed to secure some of the deadliest microbes on the planet.

Another report found that the 2009 outbreak of swine flu involved a virus, preserved in laboratories since the 1950s, that was probably released *by mistake* in the late 1970s.[28] This was an alarming discovery, since the swine flu epidemic had global consequences, with over 60 million cases of infection, roughly 270,000 hospitalizations, and more than 12,000 deaths between 2009 and 2010.[29]

Many other examples could be mentioned. The point should be clear enough, though: the risk of pestilence stems not only from terror but from error as well. Both are nontrivial risks to future human well-being, and both will likely become more acute this century. This is the anthropogenic side of a pandemic's dual etiology. In addition, disease-causing bugs still lurk in the shadows of the natural world, threatening to leap out and infect people at any moment. Although there are mechanisms associated with modern civilization that can mitigate the risk of an outbreak, there are perhaps more factors that will exacerbate it.

Notes

1. See Candida Moss, "How an Apocalyptic Plague Helped Spread Christianity," *CNN, Religion Blogs*, June 23, 2014, http://religion.blogs.cnn.com/2014/06/23/how-an-apocalyptic-plague-helped-christianity/.

2. See Pan American Health Organization, "Food Security in a Pandemic," https://www.google.com/url?sa=t&rct=j&q=&esrc=s&source=web&cd=1&ved=0CB8QFjAAahUKEwiHwpKqgpXGAhXhPowKHUqcDPQ&url=http%3A%2F%2Fwww.paho.org%2Fdisasters%2Findex.php%3Foption%3Dcom_docman%26task%3Ddoc_download%26gid%3D533%26Itemid%3D&ei=XoKAVYebF-H9sATKuLKgDw&usg=AFQjCNEuIngBP4BVNp5RhqdWrFlje1KlcQ&sig2=ZYldqMYVwC2Y4YkQcA1nfA&bvm=bv.96041959,d.cWc.

3. With possible exceptions in chapters 7, 8, and 9.

4.   Interestingly, the weaponization of contagious diseases stretches back centuries. For example, battles were fought in the Middle Ages in which the carcasses of bubonic plague victims were catapulted over castle walls to sicken the enemy. And during the Revolutionary War, George Washington came to believe that the British "intended Spreading the Small pox amongst us," describing the virus as "the greatest enemy of the Continental Army." In 1863, during the US Civil War, Confederate troops poisoned Union wells with animal carcasses, while others contaminated drinking water by driving "animals into ponds and then [shooting] them." And the "Southern sympathizer and later governor of Kentucky," attempted (and possibly succeeded) to sell clothing contaminated by smallpox to Union soldiers. More recently, the Japanese may have killed some 400,000 Chinese during World War II by tainting towns with nasty critters like anthrax, cholera, and the plague. And it's possible that the United States used biological weapons in the Korean War, although whether this happened remains contentious.

5.   For more, see the Committee on Advances in Technology and the Prevention of Their Application to Next Generation Biowarfare Threats, "Globalization, Biosecurity, and the Future of the Life Sciences," 2001, http://www.biosecurity.sandia.gov/ibtr/subpages/papersBriefings/2005-2006G/FutureLifeSciences.pdf

6.   See Eliza Strickland, "Did Craig Venter Just Create Synthetic Life? The Jury Is Decidedly Out," *Discover Magazine, 80 Beats* blog, May 20, 2010, http://blogs.discovermagazine.com/80beats/2010/05/20/did-craig-venter-just-create-synthetic-life-the-jury-is-decidedly-out/#.VBddiS5dXBc.

7.   See Michael Bernstein and Michael Woods, "J. Craig Venter, Ph.D., Describes Biofuels, Vaccines and Foods Form Made-to-Order microbes," American Chemical Society, March 2012, http://www.acs.org/content/acs/en/pressroom/newsreleases/2012/march/craig-venter-phd-describes-biofuels-vaccines-and-foods-from-made-to-order-microbes.html.

8.   See Leslie Tzeng, "Synthetic Biology: Benefits, Risks, and Regulations," *Triple Helix Online*, September 12, 2013, http://triplehelixblog.com/2013/09/synthetic-biology-benefits-risks-and-regulations/.

9.   See "Globallization, Biosecurity, and the Future of the Life Sciences," Committee on Advances in Technology and the Prevention of Their Application to Next Generation Biowarefare Threats, Development, Security, and Cooperation, 2006, p. ix.

10.   See "Biotechnology Research in an Age of Terrorism: Confronting the 'Dual-use' Dilemma," National Academies Press, 2003.

11.   See "The Darker Bioweapons Future," CIA, November 3, 2003, http://fas.org/irp/cia/product/bw1103.pdf.

12.   See Jonathan Tucker, "Could Terrorists Exploit Synthetic Biology?" *New Atlantis*, Spring 2011, http://www.thenewatlantis.com/publications/could-terrorists-exploit-synthetic-biology.

13.   See "12 Risks That Threaten Human Civilisation," Global Challenges Foundation, February 2015, p. 15.

14.   See Andrew Pollack, "Traces of Terror: The Science; Scientists Create a Live Polio Virus," *New York Times*, July 12, 2002, http://www.nytimes.com/2002/07/12/us/traces-of-terror-the-science-scientists-create-a-live-polio-virus.html.

15.   See Jonathan Tucker, "Could Terrorists Exploit Synthetic Biology?" *New Atlantis*, Spring 2011, http://www.thenewatlantis.com/docLib/20110805_TNA31Tucker.pdf.

16.   See J. van Aken, "Ethics of Reconstructing Spanish Flu: Is It Wise to Resurrect a Deadly Virus?" *Nature* 98, no. 1–2 (2007), http://www.nature.com/hdy/journal/v98/n1/full/6800911a.html.

17.   See James Randerson, "Revealed: The Lax Laws That Could Allow Assembly of Deadly Virus DNA," *Guardian*, June 14, 2006, http://www.theguardian.com/world/2006/jun/14/terrorism.topstories3.

18.   See "Ebola Virus Disease," World Health Organization, September 14, 2014, http://www.who.int/mediacentre/factsheets/fs103/en/.

19.   Although not one here considered.

20.   See "12 Risks That Threaten Human Civilisation," Global Challenges Foundation, February 2015, p. 85.

21.   See Josh Hicks, "Audit: Homeland Security 'Ill-prepared' for Potential Pandemics," *Washington Post*, *Federal Eye* blogs, September 8, 2014, http://www.washingtonpost.com/blogs/federal-eye/wp/2014/09/08/audit-homeland-security-ill-prepared-for-potential-pandemics/.

22.   See "Secretary-General's Lecture at Princeton University," United Nations, November 28, 2006, http://www.un.org/sg/statements/?nid=2330.

23.   See Zoe Schlanger, "Bioterror Lab Accidents Happen Far More Often Than We Thought," *Newsweek*, August 18, 2014, http://www.newsweek.com/bioterror-lab-accidents-happen-far-more-often-we-thought-265334.

24.   See Sabrina Tavernise and Donald McNeil Jr., "CDC Details Anthrax Scare for Scientists at Facilities," *New York Times*, June 19, 2014, http://www.

nytimes.com/2014/06/20/health/up-to-75-cdc-scientists-may-have-been-exposed-to-anthrax.html.

25.   See Donald McNeil, "CDC Closes Anthrax and Flu Labs after Accidents," *New York Times*, July 11, 2014, http://www.nytimes.com/2014/07/12/science/cdc-closes-anthrax-and-flu-labs-after-accidents.html?_r=0.

26.   See Jen Christensen, "CDC: Smallpox Found in NIH Storage Room Is Alive," *CNN*, July 11, 2014, http://www.cnn.com/2014/07/11/health/smallpox-found-nih-alive/.

27.   See Brady Dennis and Lena Sun, "FDA Found More Than Smallpox Vials in Storage Room," *Washington Post*, July 16, 2014, http://www.washingtonpost.com/national/health-science/fda-found-more-than-smallpox-vials-in-storage-room/2014/07/16/850d4b12-0d22-11e4-8341-b8072b1e7348_story.html.

28.   See Steve Connor, "Did Leak from a Laboratory Cause Swine Flu Pandemic?" *Independent*, June 30, 2009, http://www.independent.co.uk/news/science/did-leak-from-a-laboratory-cause-swine-flu-pandemic-1724448.html.

29.   See "CDC Estimates of 2009 H1N1 Influenza Cases, Hospitalizations and Deaths in the United States," CDC, http://www.cdc.gov/h1n1flu/estimates_2009_h1n1.htm.

4

Manufacturing Molecules

• • • • •

A Stairway to Heaven

Your cell phone alarm goes off, gently lulling you out of an early morning daze. You slide out of bed, shuffle into the kitchen, and make yourself a cup of coffee. Three bags of clothes are waiting by the door. After a quick shower, you pick up a taxi and drive to the nearest port, where you board a high-speed watercraft headed for the *Floating Pacific Base Station*, located in the equatorial regions of the Pacific Ocean. You check your pocket for the 5:00 p.m. *Climbing Pod* ticket you bought a few months earlier. This pod will take you on a five-day ascent straight up a 22,000-mile cable into space.

You arrive at the base station in time and drag your belongings onto the *Climbing Pod*. As it lifts you high into the atmosphere, you push your face against the window and watch the horizons of Earth slowly converge. The blue azure once above you morphs into a black canopy of space below, with the shrinking dot of Earth a splash of color amid the infinite darkness around it. Each day of travel, your weight incrementally decreases until you start floating free in microgravity. Finally, you arrive at the *Pacific Geostationary Platform*, a structure that orbits Earth at about 22,000 feet. This platform remains fixed above the *Floating Pacific Base Station* to which it's tethered. Here the last stage of your journey begins, as you board an interplanetary shuttle: you're on your way to visit a friend from college, and she lives on Mars.

A trip like this may sound improbable, if not silly. But a number of institutions and companies around the world right now are working to build fully functional *space elevators*, which would offer a far cheaper way than rockets to transfer goods and people into outer space. (Not to mention that they'd have less of a carbon footprint, as a single space shuttle launch

72

produces 28 tons of carbon dioxide.[1]) For example, NASA has funded research in this area, and a Japanese company recently claimed that it will be shuttling passengers out of Earth's atmosphere at a speed of 124 mph by 2050.[2]

It wasn't that long ago that engineers deemed such technology impossible, primarily because a space elevator would require a *huge* cable to connect the base station, most likely located in the ocean (so it can move around to avoid inclement weather), to a counterweight at the other end, orbiting "geosynchronously" above the base. No known material was strong enough to span even a fraction of this distance—that is, until the invention of *carbon nanotubes*. These are, "on an ounce-for-ounce basis . . . at least 117 times stronger than steel and 30 times stronger than Kevlar."[3] If long carbon nanotube strands were woven together to create a "ribbon," its tensile strength could be sufficient for the cable to vertically traverse Earth's atmosphere, thus making possible an elevator into space.

This is an example of how *nanotechnology* will revolutionize the future. Popularized by the nanotech pioneer Eric Drexler, the term refers to a broad range of methods, techniques, areas of research, and artifacts. The common feature is a focus on the extreme miniature—specifically, on regions of space that are a *billionth* of a meter across, or the length of about *a few atoms* placed next to each other.[4] There are many "nanoscale" products already in use today. Examples include nanoparticle-filled sunscreen lotions, synthetic bone that can replace human bone, tennis balls with nanocomposite materials enabling them to last twice as long,[5] and more energy efficient, brighter displays using "organic light-emitting diodes." In 2013, a group of engineers at Stanford University created the first computer with carbon nanotube rather than silicon transistors, which is significant because carbon nanotubes are not only incredibly strong but also an excellent semiconductor material. IBM recently announced that it will make nanotube transistors available on the market by 2020. This could make computers five times faster and much more efficient than the ones we have today.[6]

Nanotechnology promises many extraordinary benefits to humanity, including better products at a fraction of current costs. Unfortunately, it's no less dually usable than the centrifuges and test tubes discussed in previous chapters: along with the good will come a significant amount of danger. This is disconcerting because nanotechnology is, just like biotechnology and synthetic biology, developing at something like an exponential rate. For example, a paper by Mihail Roco, the founding chair of the US National Science and Technology Council's subcommittee on nanotechnology, examined the number of publications "reflecting discoveries in the area of nanotechnology," "the number of researchers and workers involved in one domain or another of nanotechnology," and "the value of products incorporating nanotechnology as the key component." All these phenomena exhibit explosive growth trends.[7]

Making matters worse, just like biotechnology and synthetic biology, nanotechnology is becoming increasingly *accessible*. In particular, an anticipated future version of nanotechnology will enable smaller and smaller groups to wield more and more power and influence over civilization. This has obvious implications for terrorism, but it also could lead to widespread geopolitical instability, as we'll explore in this chapter. The crucial point for now is that, of all the risks explored in this book, nanotechnology is the *other* nexus of dual usability, power, and accessibility—and this is why it ranks high among the most formidable threats to our future.

Manufacturing Problems

Nanotechnology comes in many forms. The vast majority of these pose the same sorts of risks as current industrial products and processes, although they may amplify such risks considerably.[8] For example, nanoscale technologies could produce new kinds of toxic materials, and nanoparticles with novel properties could harm organisms and disrupt ecosystems, thereby increasing the likelihood of environmental collapse. Nanotechnology could also facilitate the production of highly lethal pathogens and even amplify the risks associated with superintelligence by enabling the creation of better computers.[9]

The biggest problems, though, stem from a not-yet-realized type of nanotechnology called *molecular manufacturing* (MM). This would involve nanoscale machines, or "molecular assemblers," building larger objects by grabbing ahold of and repositioning individual molecules one at a time. Whereas the nanoscale technologies around today rely on phenomena like molecular self-assembly (a spontaneous process that exploits physics without any direct intervention—something that Drexler calls "Brownian assembly"), molecular manufacturing entails the creation of products *with absolute atomic precision*. This means, for example, not only that two microchips manufactured in this way would look the same in terms of their macroscopic properties but also that the atoms inside them would be arranged in an absolutely *identical* manner.

The ability to construct objects whose structures are precise down to the atom could benefit humanity immensely. Molecular manufacturing has the potential to make products like solar panels for far less money than is currently possible, and "with no net CO_2 emissions."[10] It could also help us clean up the pollutants already dumped into the oceans, ground, and atmosphere, thereby reversing the environmental degradation discussed in chapters 7 and 8.[11] And molecular assemblers would almost certainly revolutionize medicine. Imagine, for example, that it's 2075 and you've just been diagnosed with stomach cancer. If the year were 2015, your doctor would've given you two months to live. But thanks to advanced nanotechnology, there's not only a treatment available, but it's also nearly painless: a nurse simply injects a solution of *nanobots* into your bloodstream, and a short time later the tumor disappears. Because cancer cells secrete different chemicals than healthy cells, the army of nanobots in your body would have a way to identify them. Once this occurs, they could either destroy the cells or, maybe, just infiltrate them to turn off the problematic genes.[12] Other medical uses include nanobots releasing pharmaceuticals in targeted regions of the body, repairing cells, and apprehending pathogens.[13]

A more speculative use, which we'll return to in the next chapter, involves nanobots swimming through the fishbowl of cerebrospinal fluid atop your spine. In the process, they could scan the complete microstructure of your brain, wirelessly sending this information to a supercomputer capable of reconstructing the brain in 3D. If the best current theory in cognitive science about the nature of mental states is true (i.e., functionalism), then this simulated brain *in silica* would become as conscious and aware as you and I are. The result would be a cognitive clone capable of surviving the decay and death of your physical body.[14]

Furthermore, molecular manufacturing could lead to the construction of a truly game-changing device called a *nanofactory*. This would initiate a revolution because it would enable *virtually anyone to manufacture a huge range of products at almost no expense.* Just dropped your phone? No problem: print out a new one. Want some new clothes? No problem: print a new dress. Need some new shoes? A kitchen knife? A computer? A wood dresser? A piece of paper? A pen? Some makeup? New glasses? No problem: simply plug in the nanofactory and *voilà*, they're yours. One could even print out food for dinner, at least in theory, or manufacture homebuilt orbital spacecrafts in one's garage.[15] As the Foresight Institute states, such spacecraft could be "comparable to a van able to carry a family and its luggage to earth orbit," made of materials that are "50–100 times stronger and lighter than steel."[16] In this way, nanofactories could open up the space frontier by making rocket science accessible to anyone with an itch for outer space adventure. (A very helpful video showing how a nanofactory might work can be found at this link: https://www.youtube.com/watch?v=vEYN18d7gHg.)

All a nanofactory would need to work are three simple ingredients: blueprints, feedstock, and an energy source. The blueprints would consist of a file storable on a computer (which might be hooked up directly to the nanofactory), and the feedstock would consist of raw material—simple molecules like acetone or acetylene—out of which the factory would make stuff.[17] Some nanofactories could be rather large, while others, called *personal nanofactories* (PNs), might be quite compact, potentially small enough to fit on one's desk at home.

Perhaps the most significant product that a nanofactory could create are *other nanofactories*. Were this to occur, it would mark a watershed in human history: for the first time ever, a means for producing nearly any commodity would become *nonscarce*.[18] Nanofactories could even facilitate the acquisition of feedstock—one of three conditions for their operation—by manufacturing "feedstock processing plants."[19] As the Lifeboat Foundation calculates, if each nanofactory were to produce one additional nanofactory over 28 duplication cycles, it would take a mere *18 days* to produce 200 million units.[20] This is nearly the number of people in the US over the age of 18. Thus, the material needs of entire nations could be satisfied cheaply, and in a completely *self-sufficient* manner. Just envision such a world: everything society needs to function—computers, automobiles, medicines, and foodstuffs—could be manufactured locally, rather than by companies in other countries. The complex network of global trade would consequently disintegrate as states become reliant upon their own internal resources.[21]

Achieving such "radical abundance," as Drexler puts it, has clear benefits. The vicissitudes of the global economy—the constant menace of recession and other disruptions—could be neutralized. The new prosperity could be ubiquitous, sustainable, and far more reliable, limited only by energy, information, and feedstock.[22] But dissolving global trade relations could also radically destabilize the world, a situation that Drexler describes as "catastrophic success."[23] As the Global Challenges Foundation (GCF) puts it in a recent publication, "One of the greatest threats of nanotechnology is the possibility that it could result in a breakdown of trade between currently interdependent nations."[24] The problem here stems from the fact that economically interdependent countries rarely attack each other, since the risk of negative economic repercussions often outweighs the potential gains.[25] (According to Thomas Friedman's "capitalist peace theory," countries with McDonald's typically don't go to war with each other.) But if advanced nanotechnology enables countries to be entirely self-reliant, a major incentive for *avoiding* war will vanish.[26] Making matters worse, if nanotechnology were to enable faster recovery after battle, yet another deterrent to military action would be removed. Nanofactories could completely alter the cost-benefit calculus of conflict.

Personal nanofactories would also make it easier for small groups—even single individuals hidden below the surveillance horizon—to accumulate large arsenals of dangerous weaponry. An analogous technology exists right now in 3D printers ("additive manufacturing"), which could be characterized as a *very* low-tech version of the personal nanofactory.[27] The products of 3D printers, such as car parts and prosthetic limbs, are often made of plastic. (There are exceptions: some can print with metal, and others can even print *human organs*, a technology that promises faster and safer organ transplants.) Although 3D printers aren't yet a household item, they've become increasingly common since 2005, when new designs and open-source software to build them at home hit the market.

What's notable is that Second Amendment enthusiasts quickly realized that 3D printers could be used to manufacture guns in the home (or, perhaps, in one's backyard bunker). In 2012, a gun-loving libertarian named Cody Wilson founded Defense Distributed, an organization that developed a printable pistol called the "Liberator" and put the design online for free. Despite being made of plastic, the gun can fire real bullets—and consequently kill real people. Within two days of uploading the Liberator's design, the US government forced Defense Distributed to remove it from the Internet, but at that point the file had already been downloaded over 100,000 times. The same year Cody Wilson started Defense Distributed, *Wired* magazine named him one of the 15 most dangerous people in the world.

Wilson represents a sizable portion of the US population. Such individuals believe that any restriction on one's access to weapons is an assault on personal freedom. It's not hard to imagine how the proliferation of desktop nanofactories could profoundly exacerbate this situation. Instead of a pistol made of plastic, a nanofactory owner could print out an *actual* AK-47, along with its cartridges. One could even manufacture weapons far more lethal than guns, such as grenades and bombs. The arrival of nanofactories could thus bring about a period of rapid and widespread stockpiling of immensely powerful weapons not only by governments but also by militias and people like Cody Wilson. This could initiate a vicious cycle whereby those who wouldn't otherwise own guns begin manufacturing them out of fear, until everyone is armed to the teeth.

Although the government forced Cody's company to retract the Liberator's design, once information makes it to the Internet, it's virtually impossible to contain. Indeed, the Liberator file can be downloaded (as of this writing) by anyone from websites like the Pirate Bay. This suggests that future regulation of personal nanofactories may be extremely difficult to achieve. As with biotechnology and synthetic biology, the threat of citizens manufacturing weapons could push governments toward more totalitarian, surveillance-type states, motivated by the slogan "Security over Privacy." (Recall Ben Franklin's sagacious warning that those who follow this slogan deserve neither security nor freedom.)

On the international stage, the proliferation of nanofactories could initiate arms races between countries attempting to gain a strategic advantage over their rivals. Unlike the arms race of the Cold War, though, which involved exactly *two* military powers and was governed by the stable equilibrium of mutually assured destruction (MAD), a nanotech arms race with "constantly evolving arsenals" could involve multiple states, and it's not clear that the logic of MAD would apply.[28] If a state gained even a momentary advantage, for example, it could quickly annihilate its competitors and take over their territories, perhaps eventually implementing a global government, if enough countries are defeated. The aftermath of a nanotech war might be easier to clean up than that of a nuclear conflict, making an attack using nanoweapons more appealing to states. As the Center for Responsible Nanotechnology (CRN) notes, an "all-out nanotech war is probably equivalent [to a nuclear war] in the short term, but nuclear weapons also have a high long-term cost of use (fallout, contamination) that would be much lower with nanotech weapons."[29]

Furthermore, whereas "nuclear weapons cause indiscriminate destruction, nanotech weapons could be targeted," affecting specific resources or capabilities of the enemy.[30] Factors like these make an arms race involving the proliferation of "horrifically effective weapons," to quote Louis Theodore and Robert Kunz in *Nanotechnology: Environmental Implications and Solutions*, among multiple state actors highly precarious.[31] If such a race were to commence—and already, DARPA has announced a nanotech research and development program called "Atoms to Products"— the outcome could be quite different than the Cold War.

In a world full of more powerful and highly accessible guns, self-guided bullets, new aerospace materials that enable aircraft to avoid detection, embedded computers allowing for the remote activation of weapons,[32] invisibility cloaks, tiny combat vehicles the size of an insect, *and so on*, the nanotech revolution could push civilization into a precarious state of global *disequilibrium*. As the late David Jeremiah, an admiral with the Joint Chiefs of Staff, presciently stated in 1995, "Military applications of molecular manufacturing have even greater potential than nuclear weapons to radically change the balance of power."[33]

Ecology Eaters

All the scenarios mentioned above fall squarely within the bounds of technological plausibility. A more speculative risk derives from the possibility of *self-replicating autonomous nanobots*. Unlike nanofactories, which are designed to be stationary (atop one's desk, for example), nanobots would be able to move through the environment on their own. Some could, at least in theory, even be designed to draw energy from the environment and convert the matter around it into other nanobots, which would then convert more matter into more nanobots, and so on.

If such a feedback process were to commence, the entire planet could be converted into a swarm of mindlessly reproducing nanoscale machines in a rather short time. The futurist Ray Kurzweil estimates that an existential catastrophe of this sort could unfold in as little as *90 minutes*, although this claim is disputed by other experts.[34] While none of the current proposals for molecular manufacturing call for the creation of self-replicating nanobots, a nanoterrorist harboring a death wish for humanity could *intentionally design* such critters, releasing them into the biosphere to bring about the ultimate genocidal disaster. This is called the *grey goo scenario*—although being eaten alive by tiny, rapacious machines need not produce anything grey *or* gooey.

In conclusion, the promise of nanotechnology is immense. There's a good chance that our world will be transformed as the field of nanotechnology undergoes an exponential growth spurt in the coming decades. Because of this transformation, the life of our descendants 200 years from now may be as unrecognizable to us as our lives would be to a hunter-gatherer from the Paleolithic Era. But the dual usability of nanotechnology means that great dangers lurk on the horizon. This is due in part to the fact that nanotechnology is, like biotechnology and synthetic biology, expected to become both increasingly powerful *and* accessible, along the four axes outlined in chapter 1. One might, indeed, anticipate something like a *nanohacker* movement to emerge, on the model of biohacking, but focused on even lower-level manipulations of nonliving matter.

This confluence of trends is cause for anxiety because, again, the number of psychopaths in the world is far greater than the number of terrorist groups, just as the number of terrorist groups is greater than the number of warmongering states. It will, therefore, become increasingly likely that a catastrophe of genuinely unprecedented—perhaps even *eschatological*—proportions will occur without us even having known the perpetrator posed a threat. Yet attempts to contain this situation will introduce a different kind of danger: a global totalitarian state that oppressively monitors its citizens. This is the Franklinian dilemma we'll have to navigate as the twenty-first century proceeds.[35]

Notes

1.   See Prachi Patel-Predd, "A Spaceport for Treehuggers," *Discover Magazine*, November 26, 2007, http://discovermagazine.com/2007/dec/a-spaceport-for-tree-huggers.

2.   See Surojit Chatterjee, "Space Elevator That Soars 60,000 Miles into Space May Become Reality by 2050," *International Business Times*, February 21, 2012, http://www.ibtimes.com/space-elevator-soars-60000-miles-space-may-become-reality-2050-709656; and Mike Wall, "Japanese Company Aims for Space Elevator by 2050, *Space.com*, February 23, 2012, http://www.space.com/14656-japanese-space-elevator-2050-proposal.html.

3.   See "Carbon Nanotubes Twice as Strong as Once Thought," *Science Daily*, September 16, 2010, http://www.sciencedaily.com/releases/2010/09/100915140334.htm.

4. See Chris Phoenix and Mike Treder, "Nanotechnology as Global Catastrophic Risk," in *Global Catastrophic Risks*, ed. Nick Bostrom and Milan Ćirković (Oxford University Press, 2008), p. 481.

5. These balls, the Wilson Double-Core, are now the official ones used in the Davis Cup. See Robert Paull, "The Top Ten Nanotech Products of 2003," *Forbes*, December 29, 2003, http://www.forbes.com/2003/12/29/cz_jw_1229soapbox.html.

6. See Tom Simonite, "Chips Made with Nanotube Transistors, Which Could Be Five Times Faster, Should Be Ready around 2020, Says IBM," *MIT Technology Review*, July 1, 2014, http://www.technologyreview.com/news/528601/ibm-commercial-nanotube-transistors-are-coming-soon/.

7. See Mihail Roco, "The Long View of Nanotechnology Development: The National Nanotechnology Initiative at Ten Years," National Science Foundation, http://www.nsf.gov/crssprgm/nano/reports/nano2/chapter00-2.pdf.

8. See Chris Phoenix and Mike Treder, "Nanotechnology as Global Catastrophic Risk," in *Global Catastrophic Risks*, ed. Nick Bostrom and Milan Ćirković (Oxford University Press, 2008), pp. 483–484.

9. See Chris Phoenix and Mike Treder, "Nanotechnology as Global Catastrophic Risk," in *Global Catastrophic Risks*, ed. Nick Bostrom and Milan Ćirković (Oxford University Press, 2008), p. 484.

10. See Eric Drexler, *Radical Abundance: How a Revolution in Nanotechnology Will Change Civilization* (Perseus Books Group, 2013), p. 51.

11. See "Benefits of Molecular Manufacturing," Center for Reponsible Nanotechnology, accessed June 16, 2015, http://www.crnano.org/benefits.htm.

12. See Jon Weiner, "Caltech-led Team Provides Proof in Humans of RNA Interference Using Targeted Nanoparticles," Caltech, March 21, 2010, http://www.caltech.edu/article/13334.

13. See Arlington Hewes, "DARPA's New Initiate Aims to Make Nanoscale Machines a Reality," *Singularity HUB*, August 31, 2014, http://singularityhub.com/2014/08/31/darpas-new-initiative-aims-to-make-nanoscale-machines-a-reality/.

14. This remains speculative. For more, see Ray Kurzweil, *The Singularity Is Near* (Penguin Group, 2005), p. 200.

15. See Robert Frietas Jr. and Ralph Merkle, "Molecularly Precise Fabrication and Massively Parallel Assembly: The Two Keys to 21st Century Manufacturing," Molecular Assembler, October 28, 2002, http://www.molecularassembler.com/

Nanofactory/TwoKeys.htm; and Robert Freitas Jr., "Economic Impact of the Personal Nanofactory," *Nanotechnology Perceptions*, 2 (2006), http://www.rfreitas.com/Nano/NoninflationaryPN.pdf.

16. See "From Feynman to Nanofactories," Foresight Institute, accessed on June 17, 2015, http://www.foresight.org/nano/nanofactories.html. Thanks to Chris Phoenix for pointing out this option in personal communication.

17. See Chris Phoenix and Mike Treder, "Nanotechnology as Global Catastrophic Risk," in *Global Catastrophic Risks*, ed. Nick Bostrom and Milan Ćirković (Oxford University Press, 2008), pp. 485–486. Thanks to Chris Phoenix for clarifying this point for me.

18. See Chris Phoenix and Mike Treder, "Nanotechnology as Global Catastrophic Risk," in *Global Catastrophic Risks*, ed. Nick Bostrom and Milan Ćirković (Oxford University Press, 2008), p. 485.

19. See Chris Phoenix and Mike Treder, "Nanotechnology as Global Catastrophic Risk," in *Global Catastrophic Risks*, ed. Nick Bostrom and Milan Ćirković (Oxford University Press, 2008), p. 486.

20. See "First-Stage Nanoproducts and Nanoweaponry," Special Report, Lifeboat Foundation, accessed on June 17, 2015, http://lifeboat.com/ex/nanoweaponry.

21. See Chris Phoenix and Mike Treder, "Nanotechnology as Global Catastrophic Risk," in *Global Catastrophic Risks*, ed. Nick Bostrom and Milan Ćirković (Oxford University Press, 2008), p. 490.

22. On the other hand, as the Center for Responsible Nanotechnology points out, "disruption of the basis of economy is a strong possibility," and "nano-built products may be vastly overpriced relative to their cost, perpetuating unnecessary poverty." See "Dangers of Molecular Manufacturing," Center for Responsible Nanotechnology, accessed June 17, 2015, http://www.crnano.org/dangers.htm.

23. Perhaps information will still be traded, but material goods might not be. See Eric Drexler, *Radical Abundance: How a Revolution in Nanotechnology Will Change Civilization* (Perseus Books Group, 2013), p. 35.

24. See "12 Risks That Threaten Human Civilisation," Global Challenges Foundation, February 2015, p. 117.

25. Again, see Chris Phoenix and Mike Treder, "Nanotechnology as Global Catastrophic Risk," in *Global Catastrophic Risks*, ed. Nick Bostrom and Milan Ćirković (Oxford University Press, 2008), p. 490.

26. See T. McCarthy, "Molecular Nanotechnology and the World System,"

http://www.mccarthy.cx/WorldSystem/intro.htm.

27.   See Eric Drexler, *Radical Abundance: How a Revolution in Nanotechnology Will Change Civilization* (Perseus Books Group, 2013), p. 76.

28.   See Chris Phoenix and Mike Treder, "Nanotechnology as Global Catastrophic Risk," in *Global Catastrophic Risks*, ed. Nick Bostrom and Milan Ćirković (Oxford University Press, 2008), p. 490.

29.   See "Dangers of Molecular Manufacturing," Center for Responsible Nanotechnology, accessed June 17, 2015, http://www.crnano.org/dangers.htm.

30.   Again, see "Dangers of Molecular Manufacturing," Center for Responsible Nanotechnology, accessed June 17, 2015, http://www.crnano.org/dangers.htm. Note that I've changed a semicolon in this sentence to a comma.

31.   See Louis Theodore and Robert Kunz, *Nanotechnology: Environmental Implications and Solutions* (John Wiley & Sons, Inc., 2005).

32.   Again, see "Dangers of Molecular Manufacturing," Center for Responsible Nanotechnology, accessed June 17, 2015, http://www.crnano.org/dangers.htm.

33.   I borrow this quote from "Dangers of Molecular Manufacturing," Center for Responsible Nanotechnology, accessed June 17, 2015, http://www.crnano.org/dangers.htm.

34.   See Ray Kurzweil, *The Singularity Is Near* (Penguin Group, 2005), p. 425.

35.   I would highly recommend reading the chapter, cited earlier, by Chris Phoenix and Mike Treder in *Global Catastrophic Risks*. It provides probably the best overview of nanotech-related risks that I've come across.

5

Our Children Might Kill Us

• • • • •

Killer Computers

Sometimes truth is stranger than science fiction. Perhaps there's no better example of this than superintelligence. At first glance, the possibility of being murdered by a superintelligent being may sound absurd. Yet those who understand the topic consider it anything but. The famed physicist Stephen Hawking, for example, has expressed "extreme concerns" about the creation of superintelligence, while the influential theorist Eliezer Yudkowsky describes "smarter-than-human intelligence" as "probably the single most dangerous risk we face."[1] Similarly, the Oxford University neuroscientist Anders Sandberg ranks it among the top-five threats facing humanity, and the director of Oxford's FHI, Nick Bostrom, argues in his 2014 best-seller *Superintelligence* that we should recognize "doom" as the "default outcome" of a successfully engineered supermind. More recently, Hawking, Yudkowsky, Sandberg, Bostrom, and many other researchers (myself included) signed an open letter drafted by the Future of Life Institute (FLI) in 2015 that warns about the "potential pitfalls" of superintelligence. The document also notes that, as we'll explore below and in chapter 14, superintelligence has "great potential" to revolutionize the world for good, such as by "[eradicating] disease and poverty."[2]

Why are so many people nervous about superintelligence? What makes this rather bizarre scenario something worthy of concrete concern? The primary risks pertain to economic disruptions, arms races, and cognitive self-amplification. Before exploring these phenomena, though, we need to establish some basic concepts about the nature and feasibility of superintelligence. This may be, with a few exceptions, the most technical part of the book. Nonetheless, given how significant the risks associated with superintelligence are, understanding these issues will be well worth the effort.

Flavors of Brilliance

A superintelligence can take two general forms. These are distinct but not mutually exclusive, and indeed there's reason to think that one will occur along with the other. The first is what we might call *quantitative superintelligence*. It refers to a mind with the same basic capacities as the human mind, except that it can (a) organize and retain more information, and/or (b) process information much faster than the best human brains.[3]

What could a quantitative superintelligence do? Let me explain using an example that actually bears on the problem of big-picture hazards. Consider the fact that the *difference* between what individual people today know relative to what the collective whole knows is growing. On the one hand, humanity has acquired new knowledge at something like an exponential rate since the Scientific Revolution. On the other, the human brain has remained more or less fixed in its capacities, possibly for the past 30,000 years or longer. The result is an exponentially rapid divergence between the knowledge held collectively and that held individually: an *ignorance explosion*. As one scholar notes, "It was possible as recently as three hundred years ago for one highly learned individual to know everything worth knowing. By the 1940s, it was possible for an individual to know an entire field, such as psychology. Today the knowledge explosion [of the collective] makes it impossible for one person to master even a significant fraction of one small area of one discipline."[4] This seemingly paradoxical situation entails, as I've put it elsewhere, that *everyone today knows almost nothing about most things*.[5]

A quantitative superintelligence, though, could fix this situation: it could close the gap between individual and collective knowledge by overcoming the cognitive hurdles of memory (information retention) and time (processing speed). It could acquire, potentially, more and more knowledge until everything *known* by the whole at a given moment is something it *knows*. A quantitative superintelligence with the property of (b) could process and organize this knowledge faster than the human brain, perhaps by orders of magnitude, especially if its brain is made of silicon rather than neurons. Indeed, "biological neurons operate at a peak speed of about 200 Hz, a full seven orders of magnitude slower than a modern microprocessor (~2GHz)."[6] As Yudkowsky observes in a chapter of *Global Catastrophic Risks*, if the human brain were sped up a million times, "a subjective year of thinking would be accomplished for every 31 physical seconds in the outside world, and a millennium would fly by in eight-and-a-half hours."[7] This would give the sped-up mind ample time to internalize the whole of collective knowledge, and, in this way, a quantitative superintelligence with the properties of (a) and (b) could potentially eliminate relative ignorance, thereby reversing the ignorance explosion.*

The second type of superintelligence can be called *qualitative superintelligence*. It's based on the fact that there's a fundamental difference between (i) processing larger amounts of information at a faster speed, and (ii) gaining access to a whole new *library of concepts*. By analogy, no matter how hard it tries, a chipmunk will never have a thought about the stock market. The concept of a *market* is simply beyond its cognitive grasp. In this exact way, there are concepts—perhaps a huge number—that are *in principle* beyond our epistemic reach. We can't know them no matter how hard we might try. We wouldn't understand them even if God himself were to climb out of the sky and explain them to us with perfect lucidity.

There are two possibilities here. First, there may be questions about phenomena that we can ask but *never* answer. In such cases, the human mind is able to sort of *glimpse* the ideas dangling before us, yet no amount of squinting will enable us to actually see them. Possible examples include the 8 extra spatial dimensions posited by string theory and "wavelike particles that are simultaneously everywhere."[8] Indeed, as the physicist Richard Feynman once famously wrote, "I think I can safely say that nobody understands quantum mechanics." The best we can do is to "grasp" such phenomena mathematically, rather than conceptually: they remain locked away in a conceptual black box, even if we can infer them through mathematical equations or detect them in a particle collider.

Some philosophers have even suggested that the field of philosophy consists almost entirely of puzzles that are in principle insoluble, even though we can ask the relevant questions, such as What is causation? Do we have free will? What is the nature of consciousness? Is life meaningful? and so on. (If such questions become answerable, the argument goes, they'll be

* From a big-picture hazards point of view, this would be good. Far too many of us are unable to see the big picture, which is required to fully grasp our evolving existential predicament.

quickly absorbed by science.) This is not necessarily because such questions are *intrinsically hard*. Our failure to devise satisfying answers could simply be due to us not possessing the right kind of mental machinery. Attempting to construct a "constitutive theory" of consciousness, for example, might be like trying to fit a square peg into a round hole: it's not that round holes are hard to fit into, you just need the right kind of peg.

In other cases, though, it's entirely possible that we can't even ask the relevant questions. This is the situation of the chipmunk: not only is it forever ignorant of the stock market, but it's also forever ignorant *of* its ignorance of the stock market. In this way, there may be phenomena with respect to which we can't even stand in awe: we lack the requisite concept-generating mechanisms to pose questions about them in the first place.

The linguist and political activist Noam Chomsky refers to this situation as *cognitive closure*. It encompasses any instance of principled unknowability, whether or not the corresponding questions can be asked. In Chomsky's terminology, *mysteries* are puzzles that permanently lie beyond our ken, whereas *problems* are puzzles that our brains are capable of knowing (even if they're not currently known).[9] Notice, importantly, that the mystery-problem boundary is species-relative: the demarcation line for *humans* is different than it is for the *chipmunk*. This is where a qualitative superintelligence enters the picture in a huge way. A superintelligence of this sort would, by definition, involve the expansion of what is in principle *knowable* for it. The resulting mind would thus be able to think thoughts that are *qualitatively different* than those accessible to our brains—thoughts that are no less inscrutable to humans than the stock market is to the chipmunk.

With a new library of concepts, a qualitative superintelligence could acquire all sorts of novel, and powerful, capabilities in the world. It could, for example, devise new theories about the universe that no human would ever understand. It could also manipulate the physical world in ways that would completely stump even the brightest human minds. Imagine a team of chipmunk scientists trying to figure out how the voice of someone in China could emerge from a small electronic device in the US. A qualitative superintelligence would have the potential to create technologies that are *no less impenetrable* to our puny minds than phones are to rodents.

The possibility of minds that can think faster, process more information, and access thoughts with respect to which we're cognitively closed suggests a number of unique risks.[10] *Homo sapiens*—the self-described "wise man"—has become the dominant species on the planet not because of our physique, but because of our superior intellectual capacities. If we were to lose our position atop the pinnacle of superiority in this domain, the results could be unprecedentedly bad—or, alternatively, unprecedentedly good.

A Core Distinction

How could we create a superintelligent mind? As it happens, there are many possible strategies, some more promising than others. The broadest distinction here concerns the *material constitution* of the superintelligence. The "biological core" approach, as I will call it, proceeds by modifying the human brain in an effort to enhance its intelligence. This is part of a more general phenomenon called *cyborgization*, which has been occurring at least since the first human species, *Homo habilis* (the "handy man"), emerged in East Africa some 2 million years ago. The aim of the biological core approach is to create *cognitively enhanced cyborgs*. For our purposes here, we'll define a cyborg as a human whose physical constitution either (a) consists of technology to some extent (as in the case of cochlear implants, pacemakers, and even reading glasses), *or* (b) has been modified in some nontrivial way by technological means. The latter category will make more sense when we examine the possibility of genetically engineering the human brain or the method of iterative embryo selection.

In contrast, the "technological core" approach discards the messiness of biology in favor of a completely artificial substrate. The goal here is the creation not of cyborgs with enhanced cognitive capacities but of *artificial general intelligence* (AGI), where "intelligence" is defined as the ability to perceive one's environment and take actions that maximize the system's chances of success.[11] This is the standard definition in the cognitive sciences, and it aligns with the philosophical notion of *instrumental rationality*, or the ability to acquire the means necessary to achieve one's ends, whatever they are. It's also what makes Bostrom's "orthogonality thesis" tautological. According to this thesis, "more or less any level of intelligence could be combined with more or less any final goal."[12] But if intelligence is defined purely in terms of means, not ends (as Bostrom himself defines it), then *of course* it can be combined with any set of ends. We'll return to this issue in the chapter's final section.

Having established these two general approaches, let's examine the various strategies within them, beginning with the biological core approach. We'll conclude with a close look at how the creation of a superintelligence could, as Hawking suggests, be the worst thing to ever happen to humanity—if not the best.

Mindware on Wetware

A cognitively enhanced cyborg could take many forms. One possibility involves brain-boosting pharmaceuticals, or *nootropics*. This is the *take a pill, think better* strategy. It may sound futuristic, but the fact is that most people around the world today consume cognitive enhancements of some sort on a daily basis, although the effects are usually not dramatic.* For example, 54% of Americans imbibe coffee each day to enhance their alertness. And millions regularly consume supplements like *ginkgo biloba* and fish oil (omega-3 fatty acids) to boost brain functioning. In his book *Superintelligence*, Bostrom notes that nicotine chewing gum and caffeine helped him complete the project.

More potent smart drugs include Adderall and modafinil, both of which have become popular on college campuses. The latter is consumed by up to 25% of students at some top universities and has been shown to augment certain aspects of cognition, like working memory.[13] Even the military is exploring this "brain Viagra" to keep soldiers up for long periods of time without the negative side effects of sleep deprivation.

None of these chemicals, though, result in anything like superhuman abilities. At best, they offer marginal, if measurable, gains in intelligence. But the possibility remains that a far more powerful mind-altering substance will be developed in the future. If so, it may become widely used and on the societal level could have a significant impact. As Bostrom shrewdly observes, a mere 1% gain in "all around cognitive performance . . . would hardly be noticeable in a single individual, but if the 10 million scientists in the world all benefited from the drug the inventor would increase the rate of scientific progress by roughly the same amount as adding 100,000 new scientists."[14]

The invention of a pharmaceutical that could produce superintelligence in an individual, though, appears unlikely. No matter how good the fuel is in your gas tank, your car's performance can only improve so much with the

---

* This gestures at a distinction between "conventional" and "radical" cognitive enhancements. Conventional enhancements are ubiquitous in the contemporary world, whereas radical ones are still in the experimental stages.

same engine under the hood. This leads to a second strategy for supersizing the mind: upgrading the cognitive engine under your scalp. One option here is to hook the brain up to a computer, resulting in a "cybernetic" loop of information exchange between user and device.

At present, all such information exchanges are mediated by the middleman of *perception*. To acquire some information from the Internet, for example, you must load the page and then read it. A central feature of cyborgization, though, is the integration of biology and technology, organism and artifact.* In the case of cognition, this means gradually eliminating the middleman of perception so the brain can access the relevant information directly. This is one of the goals of *brain-machine interfaces* (BMIs), or devices that connect either to the brain itself (a neural implant) or to the outside of the head (to measure electrical patterns on the scalp). At the moment, BMIs are primarily *therapeutic* in nature rather than *enhancive*: they aim to compensate for disabilities rather than provide healthy individuals with new capabilities. Nonetheless, they're being developed at an extraordinary rate. Scientists have, for example, already succeeded in implanting BMIs into the brains of nonhumans. A Duke University team led by Miguel Nicolelis, for example, trained monkeys to control robotic arms *by thought alone*. Eventually, Nicolelis hopes to built robotic exoskeletons that paralyzed people can be strapped into and control entirely with their minds.† The first kick of the 2014 World Cup in Brazil was, in fact, made by a paraplegic named Juliano Pinto in a mechanical exoskeleton.

BMIs thus hold some promise for achieving superintelligence, mostly likely of the quantitative sort. But the rate of information transfer would still be limited by the bottleneck of the human brain: the results of a Google search presented directly to the brain, for example, have to be *processed* in addition to being *accessed*. (BMIs would primarily help with the latter.) It follows that, like nootropics, BMIs probably won't result in any strong forms of supersmarts, at least not in the foreseeable future.

---

* Indeed, in that direction: organism --> artifact. The logical end of this trend is that biology will one day be completely replaced by technology, leading to fully artificial beings.

† Such exoskeletons could also revolutionize war. The soldier of the future will quite literally be a killing machine.

What we need is an even more drastic change to the brain. This leads to a more promising strategy: manipulate the genes responsible for human intelligence. In other words, rather than *add* something to the brain, such as drugs or electrodes, use genetic engineering to modify the brain's *endogenous structure*. After all, the human brain isn't that big compared to the chimp's brain, and it's far smaller than the whale's brain. What makes the difference between us and them is the unique functional organization of our mental machinery.

Genetic engineering the nervous systems of other organisms has already been done. In one study, scientists overexpressed the NR2B gene, which codes for a type of receptor involved in learning and memory, in mice. The resulting transgenic rodents could "learn faster, remember longer, and outperform [their] wild-type littermates in at least six different behavioral tests."[15] The researchers named these "Doogie" mice after the fictional television character Doogie Howser, a prodigy who becomes a doctor at the age of 14. Given that mice are "model organisms" that share many biological features with humans, this study suggests it may be possible to change the receptor number in humans to boost our learning and memory abilities.

As the philosopher Mark Walker points out, we also know that specific homeobox genes, like X-Otx2 in the frog, control the morphogenesis of different regions of the brain. And recent research has found a variety of human genes, including ASPM and CDK5RAP2, that are responsible for the large size of the human brain (or, more specifically, for our high "encephalization quotient"). Thus, another possibility involves manipulating these genes to engineer humans with even bigger brains, perhaps enabled by the C-section.[16] It might, indeed, take only a small tweak to produce a mind capable of understanding advanced quantum mechanics with the same ease that normal folks understand basic arithmetic.

The final strategy that we'll consider here within the biological core approach is called *iterated embryo selection*. It involves humans not "playing God" but quite literally "playing natural selection" (as we've done for millennia with domesticated animals). The idea of iterated embryo selection is straightforward: first, collect stem cells from various donor embryos. Stem cells of the embryonic sort are unique in that they have the ability to differentiate into every type of cell found in the body: the very same embryonic stem cell could turn into a liver cell, a skin cell, or a brain cell. Once collected, these stem cells could be made to differentiate into the gametes involved in biological reproduction, namely sperm and ovum (egg) cells. They could then be fused into a single cell, the zygote, as happens during fertilization. The resulting zygotes, containing half their genes from the sperm and half from the ovum, could then have their genomes sequenced in the laboratory. Those with the most desirable genetic attributes (within that generation) would be collected and the others discarded. Stem cells would then be extracted from these embryos, converted into gametes, and the process would start over (the "iterated" part of the process).

If intellectual ability were the trait desired, and if we understood the genetic basis of intelligence well enough, then iterated embryo selection could lead to a rapid increase in the intellectual abilities of our children. Bostrom and Carl Shulman calculate that selecting 1 embryo out of 10, creating 10 more out of the 1 selected, and repeating this process 10 times could result in IQ gains of up to *130 points*. This shouldn't be surprising— after all, natural selection created *our* brains, with their superior ability to engage with and manipulate the world. Iterated embryo selection is basically like faking out nature: it involves creating a selective environment in which (we've decided that) intelligence is the most important feature for organisms to have, and then skipping the whole process of development into adulthood (until the very end, of course, at which point one of the embryos grows into an adult).

A notable feature of iterated embryo selection is that it appears to be a morally acceptable form of what might be called "eugenics."[17] (This is an extremely inflammatory word, I know, but please bear with me for a moment.) The central problem with past eugenics programs—most notably, the genocidal program implemented by the Nazis—is that they heteronomously force people to act *against their own will*. This is unacceptable in the deepest ethical sense, and it's led to some of the most horrendous crimes against humanity in history. Iterated embryo selection doesn't violate anyone's status as an autonomous moral agent. Nor does it involve beings—the embryos—capable of suffering, which is at the heart of secular ethics. (That is, morally good actions are ones that alleviate suffering and promote happiness; morally bad ones amplify pain and misery.) As the science of synthetic biology continues to advance, it may even become possible to discard the embryo entirely and simply design the desired genomes on the computer. Such genomes could then be synthesized in the laboratory and inserted into a zygote. This remains a hypothetical possibility, but given the developmental trajectory of life sciences technologies, there's reason to think it may become practicable in the near future.

To summarize, the cyborgization approach to superintelligence has some potential, mostly from genetic modifications of the human brain and embryo selection techniques. There may also be combinations of these strategies that could yield additive or even *synergistic* results. Perhaps the gains from pharmaceuticals and neural implants are limited, but combining these enhancements results in large increases in cognitive capacity. One can imagine other configurations: bigger brains plus a new drug, a high-IQ embryo selected over 10 generations plus a BMI connecting it to the Internet, and so on. In the end, though, our best chance at creating a superintelligence will probably come from the technological core approach, which trades the messiness of biological tissue for the durability and speed of computer hardware. It's to this possibility that we now turn.

Steal, Copy, Invent

A superintelligence whose material constitution is entirely nonbiological could take a number of forms. The most general distinction within this category pertains to whether the strategy imitates some aspect of nature or not. I will call those that do imitate nature *biomimetic*, and those that don't *synthetic*. Let's begin with biomimesis and then turn to the possibility of cognitive synthesis. (A diagram at the end of this section will help readers keep track of these strategies.)

What could we mimic in nature to create a superintelligent mind? There are two possibilities: either we could mimic the biological brain itself—the end product of evolution—or we could mimic the evolutionary processes that produced the brain—natural selection. The first can be accomplished with varying degrees of imitation. *Whole brain emulation*, also known as *mind uploading*, is an extreme case of outright plagiarism.[18] It involves transferring the 3D microstructure of a whole brain from someone's skull to a supercomputer, where its normal functioning could then be simulated with sufficient precision. There are two necessary conditions for this to occur. First, mind uploading requires not only incredibly powerful supercomputers (since the brain is the most complex object in the known universe) but also advanced scanning technologies. This is the technological condition. And second, it requires that a particular theory about the nature of mentality called *functionalism* is true—a philosophical condition. The best evidence to date suggests that functionalism is indeed true, and extrapolations from current trends suggest that such technologies will become available at some point in the foreseeable future. So, both conditions will likely be satisfied.

(Some background theory. According to functionalism, minds are like poison in the following respect: it doesn't matter what they're *made of*, it only matters what they *do*. In other words, their defining features are their functional properties, not their underlying material substrate. It follows that minds, no less than poisons, are "substrate independent," or "multiply realizable" by different physical systems, just as long as those systems have the right *functional organization*. One version of functionalism, called *computationalism*, offers a helpful analogy: minds are the software that runs on the "wetware" of the brain. As such, if some computer hardware were to be designed that exhibits the exact same functional organization as the biological brain, then a fully conscious, awake, living mind would emerge from it. This theory is accepted by a significant portion of philosophers of mind and cognitive scientists, which is notable because it entails that we don't have the sort of immortal souls posited by nearly all the world's religions. According to *virtually every expert* in the relevant fields, substance dualism—which posits that souls and bodies are two completely distinct entities capable of existing apart from each other—is a dead theory.)

There are already a number of major research projects, backed by billions of dollars, pursuing the goal of uploading a mind. One strategy for accomplishing this is a postmortem process called the *microtome procedure*. It involves, first, making the deceased brain sectionable by solidifying it. This can be done by freezing it "to liquid nitrogen temperatures."[19] The brain is then sliced into sections small enough for an electron microscope (or an even better future technology) to scan, one by one. The information gathered is sent to a supercomputer capable of reconstructing a molecularly precise 3D model. If the brain is subsequently simulated by the supercomputer, the lightbulb of consciousness would flash on, and the mind that was once the property of neural wetware would become an emergent property of the computer. The deceased person would suddenly *wake up*.[20]

Another strategy—one that wouldn't destroy the brain—is called the *nanotransfer procedure* (mentioned in chapter 4). Rather than cutting up the brain to scan it, it would entail injecting a swarm of nanobots into the bloodstream that could squeeze through the blood-brain barrier and send—perhaps wirelessly—detailed information about the brain's 3D microstructure to an external supercomputer.[21] This supercomputer could then simulate an exact replica of the scanned brain—a cognitive clone with all the same personality traits and memories: identical pasts, different futures.

While neither procedure would result in qualitative superintelligence, an uploaded mind would be quantitatively superior to ours, given the aforementioned speed differential between the biological brain and the circuit boards of computers. An uploaded mind could also potentially store far more information than our brains can, and this information wouldn't decay over time the way our memories do ("use it or lose it"). It follows that an uploaded mind could achieve a powerful version of *quantitative superintelligence*, by virtue of its speed and information storage potentials.

It would also be easier to enhance an uploaded mind than a biological brain. One of the problems with neural implants is the possibility of infection. But simulated brains can't get infected, nor can they bleed, and they're easy to repair if damaged. They can also be copied—essentially backed up—in case something goes wrong. Indeed, one might use a simple copy function to create a huge population of cognitive clones (perhaps from a particularly smart uploaded progenitor) that could be put to work on tasks like constructing a "theory of everything" that reconciles quantum mechanics with relativity theory, or even solving the growing problem of technogenic existential risks (a possibility we'll revisit in chapter 14). Of all the strategies here considered, whole brain emulation appears to be one of the most promising.

But the possibility of *partial brain emulation*, or *neuromorphic AI*, might be even more promising. This strategy is a less blatant form of plagiarism in which parts of the brain are simulated, rather than the brain as a whole. It seems probable that on the way to simulating an entire central nervous system, sections will have to be simulated first. These could potentially be cobbled together into various configurations, perhaps along with additional algorithms of our own design. The result would be a *patchwork mind* that only partially shares our cognitive architecture. Consequently, it could achieve a version of *qualitative superintelligence*, being able to access concepts with respect to which we are cognitively closed—as well as quantitative superintelligence, by virtue of running on computer hardware. (At the same time, a different cognitive architecture could result in it becoming *unable* to grasp some of the concepts available to us.) Because of the plausibility of neuromorphic AI and the fact that it would probably be qualitatively different than our minds, some experts identify it as the most worrisome of all the superintelligence options.

Another version of partial brain emulation is *connectionism*. This involves copying the general functional connectivity of neurons, rather than scanning parts of the brain to model in a computer. Connectionist systems consist of "artificial neurons" that process information in ways similar to the biological neurons in our brains. Thus, rather than having a fixed architecture, artificial neural networks can *change* over time— they can actually *learn* from their experiences, in a "natural" manner, by manipulating information below the level of explicit symbols. There have been many real-world successes involving neural networks. They are, for example, used in facial recognition technologies, radar systems, data mining, control systems for self-driving cars, and programs that predict whether the stock market will rise or fall, to name a few.[22] One day, perhaps, they may even be able to learn how to build better neural networks, resulting in a feedback loop to which we'll return in the next section.

Whole and partial brain emulation both focus on the *end product*: an artificial superintelligence. Another option is to copy the natural processes by which the human brain was produced. This leads to the last biomimetic strategy here considered: *artificial evolution*, which involves the implementation of natural selection in a virtual world full of virtual creatures engaged in the great Darwinian struggle for virtual survival. (Note that this is quite different than selecting embryos in the laboratory, even though iterated embryo selection does mimic natural evolution to some extent.) If we designed the selective environment such that intelligence is the feature most important for survival, it might be possible to generate creatures with some form of superintelligence. And since the course of evolution in a simulation might not track the evolution of our lineage over time, the resulting superintelligences might be not merely quantitatively but qualitatively different from us, having completely unique cognitive architectures.

Such simulations could be run at an incredibly fast rate (which is important because natural selection is a slow, transgenerational phenomenon, occurring on *geological timescales* in the natural world), and it might even be possible to skip some lengthy periods of evolution not directly relevant to the evolution of intelligence. For example, we might attempt to start the simulation after the RNA world, or perhaps around the time of the Cambrian explosion, when most major groups of animals suddenly appeared on the scene.

The idea of using the "dumb" mechanisms of natural selection to create, in teleological fashion, artificial beings with genuine forms of intelligence is at the heart of *evolutionary robotics*, a relatively new and promising field of research. Some recent experiments in this field have yielded impressive results. For example, in one conducted by Dario Floreano at the Swiss Federal Institute of Technology in Lausanne, a robot was placed in a box with a black quarter circle in one corner, which represented a "charging station." It consisted of two wheels controlled by a "brain" in the form of an artificial neural network. The brain was connected to eight light sensors that served as eyes: six in front and two in the back. The exact organization of its brain—how each neuron processed the signals sent to it—was encoded by a set of "artificial genes" that could be passed down from one generation to the next.

The robot was then given a battery life of 20 seconds. If it rolled over the charging station, its battery life would be renewed; if not, it would die. In this particular experiment, the most "evolutionarily fit" robots were those that moved around as much as possible and stayed away from obstacles. As one would expect, the first generation of robots moved around without any purpose or direction. Some rolled over the charging station purely by accident, thus gaining another 20 seconds of life. The artificial genes of the "fittest" robots were then used to create the next generation, after a few random mutations were injected into the pool, exactly as occurs in nature.

The following generation did a little better than the first. The process was repeated: the genes of the fittest robots were copied, a few random mutations were added, and another generation was created. This was done over and over again until finally, after 240 iterations of natural selection *in simulo*, a generation of robots emerged that could move around the pen without bumping into any walls. Not only that, but these robots evolved to dart toward the charging station a mere *2 seconds* before their batteries died, thereby extending their lifetimes. One could compare this level of intelligence with that of a cockroach.

Perhaps the most stunning result, though, was that the neurons inside the robots' brains became *specialized*, just as they are in biological brains. In other words, through natural selection over 240 generations, these brains evolved some cells that would activate only when the robot occupied a specific spot in the pen, and other cells that would activate only when the robot was oriented a particular way. These are exactly like the "place cells" and "head-oriented cells" found in the brains of animals, such as rats.[23] Experiments like this, incidentally, demonstrate the extraordinary power of natural selection. They offer powerful confirmation that the marvelous complexity of the biological world could have been produced through a completely lifeless, simple mechanism. Intelligence does not need to come from intelligence.

One limitation is that the only aspect of these robots that could evolve was their brains. In nature, brains evolve along with bodies. More recent experiments have thus enabled both bodies and brains to evolve together. The first study of this sort was conducted in 2011 by Josh Bongard, at the University of Vermont. He simulated a population of robots "that, like tadpoles becoming frogs, change their body forms while learning how to walk." These virtual robots had 12 moving parts and lived in a 3D virtual environment. Over the course of 5,000 simulations running in parallel for 30 hours at a time (an amount of processing that would have required 50 to 100 years if run serially), the robots evolved to spend less time in the young "tadpole" stage and more time in a mature, "four-legged" stage. By the end of the experiment, they not only acquired the ability to achieve the final goal faster, but they also gained abilities that weren't even selected for, such as maintaining their balance after being pushed to one side.[24]

Artificial evolution thus offers some hope for creating an artificial general intelligence without any direct human intervention. Perhaps the task of creating a superintelligence is so difficult that only an incredibly dumb mechanism like natural selection can solve it. After all, it created us—so why not a being even smarter?

We've now discussed whole brain emulation, patchwork minds, connectionism, and artificial evolution. Let's turn to the final possibility: creating a completely synthetic mind. This strategy is top-down rather than bottom-up. It involves the *direct programming* of software with general intellectual abilities, designed from scratch rather than from the starting point of biology, on the level of explicit symbols rather than below it (as was the case with connectionism). Historically speaking, "symbolic AI" was the approach adopted from the beginning, after a conference at Dartmouth College in 1956 founded the field. The philosopher John

Figure C. Technological Core Approach to Superintelligence (Resulting in an Artificial Intelligence)

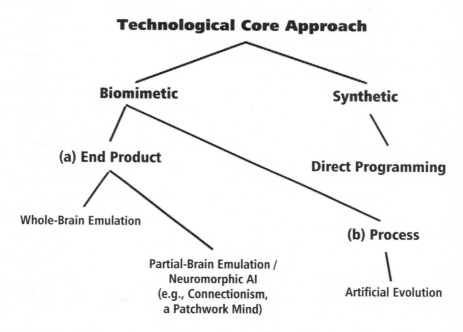

Haugeland has dubbed it GOFAI, which stands for "Good Old-Fashioned Artificial Intelligence."

While many experts were initially enthusiastic about its potential success, such hopes proved to be misguided. After several "AI winters," during which funding for artificial intelligence research dried up due to failed projects, the hope of directly programming an intelligence has reemerged in recent times. As Bostrom points out, in the long run, cognitive synthesis appears to hold the greatest possibility for producing a truly superintelligent mind, both quantitative and qualitative, since it's free of the processing constraints of biology and could acquire cognitive architectures that are radically different from ours (see figure C).

Suicide by Parricide

A superintelligence would be the most significant invention humanity has ever made, in part because it would probably be our last. On the one hand, its social, cultural, economic, political, and technological impact would almost certainly be huge. What would happen to religion after the first superintelligent mind is created? What would trading on the stock market look like? What new political systems might arise? Would superintelligence speed up the development of molecular manufacturing and other emerging technologies (perhaps some we don't even know about yet)? Could the creation of superintelligence initiate an arms race? Imagine, for example, that Russia becomes suspicious of US efforts to create a cognitive superhuman. (Incidentally, in 2013, the Defense Advanced Research Projects Agency [DARPA] initiated a project "to make computers that can teach themselves." And given the United States' history of toppling foreign governments, another country might be justified in its concern.) In response, Russia ramps up its superintelligence R&D efforts, leading to a *minds race*. As a 2015 publication by the Global Challenges Foundation notes, the competitive pressure of such a predicament could lead scientists to cut corners with respect to safety, thereby "maximising the danger" posed by superintelligence.[25]

Similar scenarios could involve nonstate actors, including terrorist groups and even single individuals. At least in theory, designing a superintelligence is the sort of task a lone wolf could accomplish in total isolation. With nefarious intentions, this individual could exploit a weaponized version of AI to inflict harm on society or trick early warning systems into indicating a barrage of nuclear missiles are headed one's way. The creation of such an AI could happen suddenly, perhaps even surprising the individual who designed it.

But why think that a superintelligence would cooperate with *anyone*? This gets at one of the most significant issues associated with the superintelligence risk, which I will call the *amity-enmity problem*. The idea is that since intelligence is defined as the ability to acquire the means necessary for some end, any level of intelligence could be combined with just about any final goal. This is Bostrom's aforementioned "orthogonality thesis," which simply points out a corollary of the standard definition of intelligence. It entails that a superintelligent mind could be just as concerned with achieving world peace, solving the arcana of the universe, and eliminating global poverty as with engaging in violent jihad or making an infinite number of paper clips.[26]

The closer a superintelligence is to our own cognitive architecture, the more likely it is, arguably, to share our goals. Conversely, the more alien its mind, the less predictable its behavior will be, from our human point of view. This gets at a crucial point: an artificial mind could have radically different ways of thinking and reasoning about the world than us. We must therefore resist the urge to *anthropomorphize* and *anthropopathize* a superintelligence, especially one with a technological core. Given an alien architecture, there's no reason whatsoever for an AI to share our cognitive traits and propensities. It may, indeed, have desires that make no sense to us. It may be interested in questions that have no importance to humanity. We might find ourselves observing it with the same perplexity of a cat watching someone sit for hours at a desk typing on a computer: *What's it doing? Who the heck knows!* As many theorists have emphasized quite vociferously, failing to resist these urges could have apocalyptically disastrous consequences.

There are, to be sure, some possible scenarios in which a super-intelligence that wants to kill us *might* be containable, although we should not underestimate how manipulative, clever, and deceitful a superintelligent being could be. For example, if building a superintelligence happened over a long period of time, in a piecemeal fashion, we might be able to anticipate a bad outcome and thus prepare an effective defense, killing it before it kills us. But if a superintelligence burst into existence—perhaps on timescales too short for even a vigilant observer to prepare—we could be taken off guard, making us vulnerable to a *superintelligence takeover*. A fast takeoff scenario is not implausible, and in fact many experts think it's more likely than the slow and moderate scenarios. Borrowing an idea from Yudkowsky, the creation of a superintelligence might be like producing the first nuclear chain reaction: it took years of incremental research to set up the necessary conditions for a nuclear reaction to begin. But then, in a moment, the reaction was initiated, and it unfolded at an exponential rate.* "The first moral," Yudkowsky writes, "is that confusing the speed of *AI research* with the speed of *a real AI once built* is like confusing the speed of physics research with the speed of nuclear reactions."[27]

The most common approach being taken by superintelligence projects might itself be more conducive to a fast takeoff scenario. The reason is that such projects focus on building a *Seed AI*, or a system that's capable of *recursive self-improvement*. The idea is simple but deep: since creating a higher intelligence is an intellectual task, an intelligence more clever

---

* Or, as Vernor Vinge put it, "The best answer to the question, 'Will computers ever be as smart as humans?' is probably 'Yes, but only briefly.'" Biological intelligence is likely a tiny stepping stone to artificial superintelligence.

than us would be better suited to accomplish it. As a result, an intelligence even *slightly* more capable than us could take over the project, improving itself until—maybe in a flash—a mind *profoundly* more capable than ours emerges. This positive feedback loop of self-amplification would initiate an *intelligence explosion* that could launch the resulting Seed AI on a history-rupturing trajectory of exponentially expanding cognitive power. While such an event—called "the Singularity" in the field of AI—could happen with a biological core being, it appears much more likely to happen with a hardware-based intelligence.

If the intelligence of a Seed AI suddenly exploded through recursive self-improvement, we could find ourselves confronted, for the first time in human history, by a mind capable of thinking *new kinds of thoughts* at a pace *much too fast* for us to keep up. A qualitative superintelligence could potentially manipulate the world in ways that no human could possibly understand: from our perspective, we'd see things happening but have no way of explaining how or why. If such a being happened to favor enmity over amity—either because we failed to implement our design correctly, or because we failed to come up with the right design in the first place—it could wipe out our species with the ease of a child stomping on a spider.[28]

Even worse, an unfriendly superintelligence might conceivably derive pleasure from torturing and killing us one by one. A humanicidal sadist with superpowers could bring about an existential catastrophe of the worst variety: tremendous suffering followed by annihilation. One need only look at how our species has treated the rest of the biosphere—a topic covered in chapter 7—to realize the plausibility of such a scenario. Indeed, there appear to be far more ways for a superintelligence to favor enmity rather than amity, which is in part why Bostrom argues that we should acknowledge the "default outcome" of an engineered supermind to be "doom."

These scenarios all fall within the category of terror. Interestingly, there's also the possibility of error, something that hasn't been discussed much in the literature. Imagine that we succeed in creating a friendly superintelligence—a being who genuinely cares about our prosperity and happiness. Here we might propose something akin to the orthogonality thesis, except focusing on intelligence and fallibility rather than intelligence and motivations. Call it the *orthogonality thesis of fallibility*. It states that higher levels of intelligence aren't necessarily correlated with the avoidance of mistakes. Albert Einstein is entirely capable of tripping on his shoelaces, just like the village idiot. Given the extraordinary powers a superintelligence would wield in the world, it might take only a *single* mistake to snuff out the human race. Thus, in the end, we could be accidentally nudged over the cliff of extinction rather than intentionally pushed. Call this the *clumsy fingers problem.*‍* What can we say? It's only superhuman.

Incidentally, superintelligence is the only technology here considered that not only has error and terror options but also yields a third category located somewhere between these extremes. That is, it could be that our superintelligent children are neither benevolent nor belligerent. They might just be *apathetic*, not giving a damn either way about whether we prosper or perish. If such a superintelligence were to want more raw materials, for example, it might harvest the molecules in our bodies. Too bad for us: humanity could be slaughtered with the indifference that someone mowing a lawn shows to the grass under her feet. Or perhaps it wants to generate more energy from the sun, and so it converts the entire surface of the earth into a vast solar array.[29] Again, too bad for us: we could end up being the collateral damage of a goal that has nothing to do with humanity, or at least not directly. Call this the *indifference problem.*[30]

But *how* exactly could an unfriendly or indifferent (or merely clumsy) superintelligence bring about such catastrophes? This is less than obvious in the case of *artificial* minds, since they're bound to the hardware of a computer rather than being autonomous agents freely moving through the world (like a cognitive cyborg). Nonetheless, it turns out that there are several plausible ways for an artificial superintelligence to manipulate the external world. Reflect for a moment on how much contemporary civilization depends upon the Internet. If a superintelligence were to gain access to the Internet it could—being better at computer hacking than any human—potentially seize control of "robotic manipulators and automated laboratories," access all sorts of personal information, steal identities, "acquire financial assets from online transactions and use them to purchase

---

* This is contra Bostrom, who writes that "being generally more capable than humans, a superintelligence would be less likely to make mistakes, and more likely to recognize when precautions are needed, and to implement precautions competently." I think this overstates the connection between intelligence and accident avoidance. Are humans less prone to make mistakes than chimpanzees? Maybe not. After all, global warming and biodiversity loss are human-caused phenomena. The fact is that we've made a number of significant mistakes that creatures below us on the intelligence scale haven't.

services and influence," use biological and chemical weapons on urban populations, control drones, and maybe even launch nuclear weapons. If nanofactories exist at the time—and if not, the superintelligence might simply invent them—the malicious mind could use them to produce "nerve gas or target-seeking mosquito-like robots [that] might then burgeon forth simultaneously from every square meter of the globe."[31]

In an effort to obviate such catastrophes, scientists might try to keep the superintelligence contained on a computer that's *not* connected to the Internet. But we shouldn't underestimate the degree to which a being vastly smarter than us could be manipulative—indeed, outright *charming*. It might be able to talk us into hooking it up to the Internet with promises of great rewards, maybe ones of a personal nature, as in: "Dr. Smith, I heard that your mother has Parkinson's disease. Well, as it happens, I've just figured out a way to cure this disease, and even reverse the damage that's already been done to her brain. I'd really love to help. If you let me out, I promise that helping your mom will be the first thing I do." Just *one* misstep here could open Pandora's box, pushing us past the Rubicon.

Finally, it could be possible for a superintelligence to escape the rigidity of infrastructure and acquire a physical body of sorts. The result would be an *android*, possibly like the one in *Terminator*. While this scenario may be more captivating to the imagination than scenarios in which the superintelligence remains more akin to a ghost in the machine, it is generally considered less probable.

Having now outlined (and underlined) the immense risks associated with the creation of greater-than-human intelligence, it's worth mentioning the potential benefits. The fact is that if we were to create a superintelligence that favors amity over enmity and that manages to avoid making a "stupid mistake" of apocalyptic proportions, then the human condition could be improved beyond our wildest dreams. A superintelligence that cared about us could help us maximize economic growth, eliminate world poverty, implement sustainable peace, develop life-extension technologies, and even mitigate the many other existential risks discussed in this book. It could inaugurate the colonization of space, and perhaps even lead to the cosmos "waking up," to quote the futurist Ray Kurzweil, as "the 'dumb' matter and mechanisms of the universe [are] transformed into exquisitely sublime forms of intelligence."[32] A superintelligence could, potentially, make life *super-awesome*. We'll return to this possibility in the final chapter of the book.

Notes

1. As one article puts it, paraphrasing Hawking. See Carina Kolodny, "Stephen Hawking Is Terrified of Artificial Intelligence," *Huffington Post*, May, 5, 2014, http://www.huffingtonpost.com/2014/05/05/stephen-hawking-artificial-intelligence_n_5267481.html. See also Eleizer Yudkowsky, "Artificial Intelligence as a Positive and Negative factor in Global Risk," in Global Catastrophic Risks, ed. Nick Bostrom and Milan Ćirković (Oxford University Press, 2008), p. 342.

2. See "Research Priorities for Robust and Beneficial Artificial Intelligence: an Open Letter," Future of Life Institute, accessed June 25, 2015, http://futureoflife.org/misc/open_letter.

3. For more, see my pseudonymous article Philippe Verdoux, "Emerging Technologies and the Future of Philosophy," *Metaphilosophy* 42, no. 5 (2011): pp. 682–707. This also parallels Bostrom's definition of "weak superintelligence" in Nick Bostrom, "How Long Before Superintelligence," *Linguistic and Philosophical Investigations* 5, no. 1 (2006): pp. 11–30.

4. See Gregg Jacobs, *The Ancestral Mind: Reclaim the Power* (Viking Adult, 2003), p. 22.

5. See Phil Torres, "Why 'Why Transhumanism Won't Work' Won't Work," IEET, June 24, 2010, http://ieet.org/index.php/IEET/more/verdoux20100624.

6. See Nick Bostrom, *Superintelligence: Paths, Dangers, Strategies* (Oxford University Press, 2014), p. 59.

7. See Eleizer Yudkowsky, "Artificial Intelligence as a Positive and Negative Factor in Global Risk," in *Global Catastrophic Risks*, ed. Nick Bostrom and Milan Ćirković (Oxford University Press, 2008), p. 331.

8. See Celeste Biever, "Quantum Mechanics Difficult to Grasp? Too Bad," *New Scientist*, November 2, 2011, http://www.newscientist.com/article/mg21228372.500-quantum-mechanics-difficult-to-grasp-too-bad.html.

9. The theologian Gabriel Marcel has also used these terms. See Daniel Migliore, *Faith Seeking Understanding: An Introduction to Christian Theology* (Wm. B. Eerdmans Publishing Co., 2014), pp. 3–4.

10. It also, at least in theory, opens up the possibility of new forms of historical methodologies. For example, I mentioned above that the universe is deterministic. If this is the case, then all events past and future have been fixed by the laws of nature. This means that if you know these laws perfectly, and if you had perfect knowledge of the microstructure of the universe at any given time, $t1$, then you could have perfect knowledge of any future event at $t2$. But you could *also* have knowledge of any past event at $t0$. This could, potentially (and within limits),

revolutionize the field of history, not to mention criminology, paleontology, and all other backward-in-time-looking areas of inquiry. With sufficient knowledge of a region of the causal structure of the universe, viz., a timeslice defined by the light cone of a prior event of interest (the "explanandum"), you could potentially use the laws of nature to deduce exactly what happened during that event, say, Hitler's final moments, or what Bush and Cheney talked about before the invasion of Iraq. There could thus be moments that you take to be completely private right now (e.g., private conversations between husband and wife, criminals, academics, not to mention intimate moments) that could someday be observed by other beings—say, a superintelligence with the cognitive abilities necessary to acquire the requisite knowledge.

11.   This is paraphrased from "12 Risks That Threaten Human Civilisation," Global Challenges Foundation, February 2015, p. 121.

12.   See Nick Bostrom, "The Superintelligent Will: Motivation and Instrumental Rationality in Advanced Artificial Agents," *Minds and Machines* 22, no. 2 (2012).

13.   See Catherine Nixey, "Are 'Smart Drugs' Safe for Students?" *Guardian*, April 5, 2010, http://www.theguardian.com/education/2010/apr/06/students-drugs-modafinil-ritalin; and Steve Bird, "The Dangers for Students Addicted to Brain Viagra: Drugs Claimed to Boost Your Intellect Are Sweeping Universities —but at What Cost?" *Daily Mail*, October 10, 2013, http://www.dailymail.co.uk/health/article-2451586/More-students-turning-cognitive-enhancing-drug-Modafinil-hope-boosting-grades-job-prospects.html.

14.   See *Nature* Podcast, January 31, 2008, http://www.nature.com/nature/podcast/v451/n7178/nature-2008-01-31.html.

15.   See George Siegel et al., *Basic Neurochemistry: Molecular, Cellular and Medical Aspects* (Elsevier Academic Press, 2006), p. 867.

16.   This is taken virtually *ad verbum* from my paper Philippe Verdoux, "Emerging Technologies and the Future of Philosophy," *Metaphilosophy* 42, no. 5 (2011): pp. 682–707.

17.   For more, see Robert Sparrow, "*In Vitro* Eugenics," *Journal of Medical Ethics* (2013), doi:10.1136/medethics-2012-101200; and Arthur Caplan, Glenn McGee, and David Magnus, "What Is Immoral about Eugenics?" *BMJ* (1999), http://www.bmj.com/content/319/7220/1284.

18.   See Nick Bostrom, *Superintelligence: Paths, Dangers, Strategies* (Oxford University Press, 2014), p. 30.

19.   See my paper Philippe Verdoux, "Emerging Technologies and the Future of Philosophy," *Metaphilosophy*, 42, no. 5 (2011): pp. 682–707.

20.   This is sloppy language. There is a difference between *minds* and *persons*, between *consciousness* and the *self*. Despite what some singularitarians seem to think, mind uploading is *not the same* as self uploading. The reason is that minds can be duplicated, but selves can't. Uploading would enable duplication; therefore, uploading the self wouldn't be possible, but uploading one's consciousness would.

21.   See Ray Kurzweil, *The Singularity Is Near* (Penguin Group, 2005), pp. 163–164.

22.   See Daniel Shiffman, "Neural Networks," Nature of Code, accessed on June 17, 2015, http://natureofcode.com/book/chapter-10-neural-networks/.

23.   The last several paragraphs borrow generously, and in some places *identically*, from my book *A Crisis of Faith* (Dangerous Little Books, 2012), pp. 38–42.

24.   See Joshua Brown, "For Robust Robots, Let Them Be Babies First," Vermont Advanced Computing Core, January 20, 2011, http://www.uvm.edu/~vacc/?Page=news&storyID=11482&category=vacc.

25.   See "12 Risks That Threaten Human Civilisation," Global Challenges Foundation, February 2015, p. 125.

26.   See Nick Bostrom, "The Superintelligent Will: Motivation and Instrumental Rationality in Advanced Artificial Agents," *Minds and Machines* (2012), http://www.nickbostrom.com/superintelligentwill.pdf.

27.   See Eleizer Yudkowsky, "Artificial Intelligence as a positive and negative factor in global risk," in *Global Catastrophic Risks*, ed. Nick Bostrom and Milan Ćirković (Oxford University Press, 2008), p. 325.

28.   See Eleizer Yudkowsky, "Artificial Intelligence as a Positive and Negative Factor in Global Risk," in *Global Catastrophic Risks*, ed. Nick Bostrom and Milan Ćirković (Oxford University Press, 2008), pp. 318–323.

29.   This possibility is mentioned by Nick Bostrom, *Superintelligence: Paths, Dangers, Strategies* (Oxford University Press, 2014), p. 97.

30.   The concept of "unfriendly AI" typically subsumes both the amity-enmity and indifference problems, although they're not termed as such.

31.   See Nick Bostrom, *Superintelligence: Paths, Dangers, Strategies* (Oxford University Press, 2014), pp. 96–104.

32.   See Ray Kurzweil, *The Singularity Is Near* (Penguin Group, 2005), p. 21.

6

Our Parents Might Kill Us

• • • • •

Sentient Simulants

There are two types of existence an artificial intelligence could have: on the one hand, it could live *in our world*. The previous chapter was primarily concerned with scenarios in which a superintelligence occupies the same universe that we do: this strange theater of becoming that we call reality. On the other hand, an artificial intelligence could live in a purely *simulated world*. This was the case with artificial evolution, although the point of artificial evolution, in the earlier context, was to produce a superintelligence that could ultimately be brought out of the simulation and into the world of humans.

But it's entirely possible that we create a simulated universe in which the simulated beings—*simulants* no less sentient than we are, assuming the truth of functionalism—live their entire lives. One can imagine reasons we might do this. Perhaps we simulate universes for our entertainment, rather like how the world of Truman Burbank in *The Truman Show* was built to entertain fans of reality TV. After all, we already have "life simulation" games like *The Sims* that have proven to be wildly popular. These games are getting more and more realistic, the logical terminus of this trend being, perhaps, the construction of worlds so detailed that the persons inside are (i) capable of having conscious experiences, and (ii) completely convinced that the universe in which they're enveloped is "real."

Or maybe we simulate universes for educational purposes. One can imagine a middle school project like this: "Go home tonight and run three simulations of human evolution; in each, change the selective environment slightly to see how it affects the morphology of our ancestors. Discuss your results in a one-page paper." More sophisticated versions of this might involve science departments running detailed simulations to test hypotheses about, for example, whether evolution inevitably leads to greater complexity, or the extent to which natural selection shapes our phenotypes relative to other evolutionary mechanisms, like genetic drift. In a seminal 2003 paper called "Are You Living in a Computer Simulation?" Nick Bostrom focuses on the possibility of "*ancestor-simulations*," which would aim to "simulate the entire mental history of humankind."[1] An even more interesting option, proposed by Alexey Turchin, is that future posthumans run "doomsday simulations" to study the ways a red-dot or black-dot catastrophe might materialize.[2] The knowledge gleaned from such simulations could help the simulators avoid a catastrophe in their own world.

The point is that *if* the opportunity arises—and it probably will, *ceteris paribus*, given Moore's Law—there's reason to think that future humans will create one or more simulated universes. Some of these might be run over and over again, while others might be run in parallel. Some could be started from the very beginning of cosmic evolution (the Big Bang), while others might be started in the middle of history, *à la* the Omphalos hypothesis championed by some creationists. Furthermore, since computer software is a functional kind in exactly the same sense that poison and minds are, it would even be possible for people inside simulations to run their own simulations, thereby producing a hierarchy of nested universes, stacked like matryoshka dolls, with one inside another.

This being said, if future humans were to run a large number of simulations, a very curious thing would follow: *it would imply that we ourselves are probably in a computer simulation.* Why? Imagine for a moment that you can take a step back and look at *all* the universes that exist at a given moment, as it were, "sideways on." You then randomly select an individual from a random universe and ask, "Is he or she a simulant?" Since the ratio of simulants to nonsimulants will be skewed toward the former, the randomly selected individual will, statistically speaking, most likely be a simulant. The more simulations we run, the more true this will be: in the case of billions of simulations, for example, we should say that the random individual will *almost certainly* be a simulant—again, for purely statistical reasons.

You keep doing this over and over again, randomly selecting one person after another, and find that nearly everyone is a simulant. If you were placing bets with a friend that the next person selected will be a simulant, you would win almost all the time.[3] But then something rather freaky happens: you select a random person and *it turns out to be you*. In this case, how should you answer the above question? Are you a simulant? If you accept (as Bostrom puts it) a "bland" version of the Principle of Indifference, then in the absence of any independent reason for thinking you're one of the privileged few perched atop the pinnacle of reality, you should answer the same as before: "I'm almost certainly a simulant."

It's for this statistical reason that running large numbers of simulations in the future will constitute empirical evidence that we ourselves are probably in a simulation. Bostrom calls this the *Simulation Hypothesis*, and he assigns a "less than 50% probability" to its truth—a significant percentage, to be sure, given its metaphysical peculiarity. As of this writing, no one has proposed a convincing refutation of Bostrom's argument, so there's no particular reason for thinking it's false (pretheoretic intuitions aside).

So Many Ways to Die[4]

The thought of being in a computer simulation might evoke a vague sense of digital claustrophobia, a feeling of unease aroused by the idea of participating, without one's consent, in a strange kind of voyeurism. In terms of eschatology, living in a simulation introduces a number of interesting apocalyptic possibilities, most of which pertain to *our universe getting shut down*. In the blink of an eye, without any warning, the universe could vanish into an oblivion. This leads to a perplexing eschatological question: What might cause our parents, so to speak, to turn off our simulation and go to bed? What could we do to ensure that our species persists into the future? Attempting to answer such queries takes us *deep* into the realm of fanciful speculation, but exploring the possible answers could nonetheless be productive. Perhaps we'll even hit on the right answer!

One possibility that's been discussed in the literature is that our parents simply get bored with us. Thus, maybe we need a war to keep them interested. If this were the case, then a nuclear holocaust could, paradoxically, actually make our continued survival *more likely*. The same could be said about a global pandemic and nanotech disaster. It might even be, as Turchin speculates above, that our universe is *designed* to undergo great calamities and cataclysms: we could be living in a doomsday simulation in which human civilization is, in some sense, *supposed* to self-destruct.

Or, as others have suggested, perhaps coming to believe that we're enclosed in a virtual universe leads our parents to give us a special reward.

Indeed, there are eschatological possibilities that are wholly *desirable*—not everything is gloom and doom. (As Bostrom notes, if we're in a simulation, then "an afterlife would be a real possibility.") A team of scientists at the University of Bonn, Germany, recently claimed that there might be ways of *scientifically* establishing that we're in a simulation.[5] But discovering our true metaphysical status could also "ruin" the simulation experiment in which we're involuntary participants. The fact is that knowing (or even merely suspecting) that you're being watched can affect your behavior in all sorts of ways, a phenomenon called the Hawthorne effect. Perhaps our simulators want us to "act natural" and wondering if we're simulants interferes with this. So, maybe this chapter of the book shouldn't exist. My apologies if it precipitates an existential catastrophe.

One could even imagine overpopulation making a simulation shutdown more probable. Why? Because our brains are computational objects more complex than anything else in the universe, and simulating large numbers of them might be too costly for the simulators. The same logic applies to us simulating universes once *we* get the necessary computational resources, since costs are inherited upward in the simulation stack. As Bostrom writes, "Simulating even a single posthuman civilization might be prohibitively expensive. If so, then we should expect our simulation to be terminated when we are about to become posthuman."[6]

But perhaps this is wrong. It could be that future computing technologies give us far more power than we currently anticipate, thus making a large number of stacked universes both technologically and financially feasible.[7] Even more, as far as I can tell, the very same statistical argument that makes the Simulation Hypothesis plausible also implies the existence of a multileveled simulation hierarchy. The argument essentially states that us becoming *simulators* gives us reason to believe that we're probably *simulants*. Since the existence of a simulant entails the existence of a simulator, this means that there exists a level "above" ours (occupied by our simulators), in addition to the level "below" (occupied by our simulants). Immediately, then, the Simulation Hypothesis implies at least *three* levels of reality.

We can now ask the same question as above, but apply it specifically to our simulators: "Are they simulants?" As before, the answer will be that they almost certainly are. This consequently implies a level above them, with respect to which we can ask the question again: "Are *they* simulants?" Once more, the answer will be the same. At some point we'll have to hit the top level of Ultimate Reality, but not before spanning a potentially vast number of embedded simulations. Furthermore, there's reason to believe that the structure of this hierarchy will be pyramidal—that is, thicker at the bottom than at the top. Why? Because each nested simulation will have the potential to spawn billions of additional simulations below it. It follows that a randomly selected universe will more likely come from the bottom, where most of the simulations have accumulated.

Tying this all together, then, the claim is that if we run large numbers of simulations in the future, (i) we'll almost certainly be in a simulation, (ii) a large stack of nested universes will almost certainly exist, and (iii) our universe will more likely be located at the bottom than the top (for statistical reasons). This may appear to be nothing more than philosophical acrobatics, and indeed it could be completely wrong. But there don't appear to be any obvious flaws. The eschatological point is that *the more simulations there are above us, the more possibilities there are for catastrophe.* Just as costs are inherited upward in a simulation stack, annihilation is inherited downward. This means that if even a *single* upstream universe were to collapse, so would ours; if even a *single* higher-level simulation were to get shut down, then ours would too.

The probability of such an existential catastrophe—call it *death by transitivity*—could be vastly high if the hierarchy is vastly tall and if we're located near the bottom. Maybe the simulation 10 levels above us fails to keep *its* simulators interested, for example, and as a result it gets turned off. This would entail death for us. Or maybe a civilization 20 levels above plunges into an Armageddon-like war and the university building in which our simulation stack is housed gets bombed. This too would entail the end of humanity. *Or*, perhaps a simulation 1,000 levels above gets turned off because the science project it was a part of loses funding, someone accidentally trips over the power cord, or a cup of coffee gets spilled on the computer's electronics. There may be a veritable *googolplex* of ways our simulation could get transitively terminated, with these opportunities piling up as one moves down the hierarchy.

In conclusion, life in a simulation would bring with it certain hazards, but life in a simulation stack could be downright deadly. If we run large numbers of simulations in the future, there might be reason for believing the probability of annihilation is close to 1. Recall the observation selection effect from chapter 1, which states that, given the nature of red-dot catastrophes, we can't project a history of survival into the future. Applying this to the present case, the likelihood of our simulation getting shut down could be astronomically high; we could be on the verge of catastrophe at every moment. Just because we've made it this far doesn't mean we're safe.[8]

Notes

1. See Nick Bostrom, "Are You Living in a Computer Simulation?" *Philosophical Quarterly* 53, no. 211 (2003): pp. 243–255.

2. Personal communication.

3. See Nick Bostrom, "Are You Living in a Computer Simulation?" *Philosophical Quarterly* 53, no. 211 (2003): pp. 243–255.

4. I borrow this phrase from a song written by Stuart Robinson of the indie-folk band Bombadil, called "So Many Ways to Die," https://www.youtube.com/watch?v=_U4ybU1KVcQ.

5. See George Dvorsky, "Physicists Say There May Be a Way to Prove That We Live in a Computer Simulation," *io9*, October, 10, 2012, http://io9.com/5950543/physicists-say-there-may-be-a-way-to-prove-that-we-live-in-a-computer-simulation.

6. See Nick Bostrom, "Are You Living in a Computer Simulation?" *Philosophical Quarterly* 53, no. 211 (2003): pp. 243–255.

7. Along these lines, the philosopher Mark Walker writes in a brilliant article on transhumanism and philosophy that "perhaps [a being far more intelligent than us] might smile at our claim that the speed of light is a fundamental physical law in the same way we smile at Lord Kelvin's claim that heavier than air flying machines are impossible." Future computing could employ phenomena that we are, at present, entirely unaware of, and which make possible giant stacks of nested simulations. Or, alternatively, we could be a simulation in a universe whose physical constants and laws are such that the computational cost of simulation stacks is far lower than it is in our universe. See Mark Walker, "Prolegomena to Any Future Philosophy," *Journal of Evolution and Technology* 10 (2002), http://www.jetpress.org/volume10/prolegomena.html.

8. Some of this chapter was taken *ad verbum* from Phil Torres, "Why Running Simulations May Mean the End Is Near," IEET, November 3, 2014, http://ieet.org/index.php/IEET/more/torres20141103.

7

Dinosaurs and Dodos

· · · · ·

Pruning the Tree of Life

In the first week of December 2012, over 24,000 scientists descended upon San Francisco for a yearly conference hosted by the American Geophysical Union. Influential figures from around the world shared their work on subjects ranging from geothermal systems and ocean sustainability to atmospheric electricity in thunderstorms. Ira Flatow, the host of NPR's *Science Friday*, and Dr. Subra Suresh, director of the National Science Foundation, gave keynote addresses.[1] The man behind *Titanic* and *Avatar*, James Cameron, was also in attendance. The story that got the most international press, though, had nothing to do with these high-profile attendees. It came from a relatively unknown geophysics researcher and environmental activist from the University of San Diego named Brad Werner. With short, pink hair, Werner delivered a mostly technical presentation with an unambiguously provocative title: "Is Earth F**ked?" His answer was in the affirmative: "more or less," he told a reporter for *i09* afterward.[2] What's shocking about this incident isn't that Werner believes the planet is nearing disaster but that he openly voiced an opinion, using attention-grabbing language, held by many contemporary scientists who study biology, climatology, ecology, and environmental science.

About 4.5 billion years ago, the earth formed from a giant cloud of dust swirling around the sun: the *solar nebula*. About 1 billion years later, the first single-celled organisms emerged from, perhaps, the puddles that form in rock concavities along the ocean's shores, or maybe the areas surrounding deep-sea hydrothermal vents. (Darwin himself speculated that they may have evolved in "some warm little pond.") Across geological timescales, these primordial beings morphed into an incredible range of diverse and disparate creatures. Milestones of evolution include an organism called *cyanobacteria* that polluted the atmosphere, inducing possibly "the greatest environmental catastrophe in the history of our planet"—one that also paved the way for terrestrial life.[3] The Great Oxygenation Event, as it's called, was followed by endosymbiosis (one cell—the mitochondrion—crawling inside a bigger cell), sexual reproduction, multicellularity, the Cambrian explosion, the colonization of land by plants, the colonization of land by animals, and eventually the appearance of a big-brained primate with opposable thumbs and an upright posture.

As the drama of life unfolded, the overwhelming majority of species disappeared. The best estimates to date are that the biological kinds alive today—about 8.7 million in total, according to one study[4]—constitute only 0.1% of all the creatures that have ever populated the earth. Put differently, about 99.9% of the species that were once extant are now extinct. It's worth noting that extinction doesn't necessarily imply a sad story. Of course, there are cases in which a species' lineage terminates. This is what happened with the dodo, due to human activity on the island of Mauritius. In other cases, though, a species can go extinct by evolving into one or more subsequent species, as when a species X disappears because it morphs into two new species, Y and Z. The point is that, however it happens, extinction is the rule rather than the exception—something to keep in mind as one reads this book.

The frequency at which species naturally disappear from the planet is called the "background rate of extinction." Every now and then this rate undergoes a radical shift upward and large numbers of species die out in clusters, over relatively short periods of time. These are called *mass extinction events*, and experts generally agree that a total of five have occurred since the dawn of life.

The most recent mass extinction, called the Cretaceous-Paleogene event, took place about 65 million years ago after a large comet (according to recent studies) hurled through space and crashed into the Yucatán Peninsula, in southeastern Mexico. This caused a massive chunk of Earth, in the form of dust, to be propelled into the atmosphere, initiating an *impact winter* (discussed further in chapter 9). We know this happened because comets contain a rare, on Earth, element called iridium, and in layers of Earth's crust dating back 65 million years *all around the world* we find none other than this element in unusual abundance: a smoking gun. We also find evidence that about 75% of the species on Earth disappeared around this time. This includes, most famously, the dinosaurs, although few people know that *some* dinosaurs managed to survive the chilly holocaust that followed the impact. These evolved into modern-day birds, which are technically classified as "avian" (rather than "nonavian") dinosaurs. So, next time you eat a chicken sandwich, you could refer to it just as accurately as a *dinosaur sandwich*.

The Holocene Extinction Event

The biography of Earth-originating life is marked by periods of widespread tragedy, in which huge numbers of organisms bit the dust. Unfortunately, though, I may have misspoken earlier when I said that the Cretaceous-Paleogene event was the most recent mass extinction. As I write this sentence, the evidence suggests we're in the early stages of the *sixth* mass extinction in life's 3.5 billion year history: the Holocene extinction event. The term "Holocene" refers to our current geological epoch, which began about 12,000 years ago, although some scientists have recently argued that humanity's planetary impact since the Industrial Revolution has been so profound that we should recognize a new geological epoch: the Anthropocene.

While this term has not yet been adopted by the scientific community as a whole, there are strong reasons for taking it seriously. For example, in the last few centuries, the frequency of extinctions has increased up to *10,000 times* the normal background rate, meaning that, statistically speaking, there are literally dozens of species dying out every day.[5] The result has been a significant loss of *biodiversity*, a measure of the "variability among living organisms from all sources including . . . terrestrial, marine and other aquatic ecosystems and the ecological complexes of which they are part."[6] Biodiversity is a broad concept that covers not only the number of species alive at a given time—which is what matters with respect to mass extinctions—but the diversity of ecosystems and individuals within species.

According to one of the most comprehensive studies to date, the *3rd Global Biodiversity Outlook* (GBO-3) report published in 2010, the total population of vertebrates—a category that includes mammals, birds, reptiles, sharks, rays, and amphibians—living within the tropics fell by an unbelievable 59% from 1970 to 2006. More specifically, the report found that vertebrates in freshwater environments declined by 41%, farmland birds in Europe declined by 50% since 1980, birds in North America declined by 40% between 1968 and 2003, and about 25% of all plant species—the foundation of the food chains upon which our survival depends—are currently "threatened with extinction."[7]

Another major study, the 2014 *Living Planet Report*, makes similarly alarming claims. By measuring "more than 10,000 representative populations of mammals, birds, reptiles, amphibians and fish," it found that the global population of vertebrates between 1970 and 2010 declined by a whopping 52%. On its count, freshwater species waned by a huge 76%, marine and terrestrial vertebrates each lost 39%, and populations in protected areas have dropped by 18% since 1970. The report concludes that we would "need 1.5 Earths to meet the demands we currently make on nature," which have steadily risen since the Industrial Revolution.[8]

Other studies have found that 48% of all primates, 68% of all plant species evaluated by the Center for Biological Diversity, and about 50% of turtles and tortoises are threatened with extinction.[9] Amphibian populations are especially important to track because amphibians are *ecological indicators*, or organisms particularly sensitive to perturbations in the environment. Just as canaries are more likely to be affected by carbon monoxide and methane than human beings—which is why coal miners would carry them underground—amphibians are more likely to die out when the environment is in distress. A sudden decrease in their populations is thus a harbinger of ecosystem collapse. And according to the GBO-3, "42% of all amphibian species . . . are declining in population."[10] In other words, the canary is sick: it's time to evacuate the mine.

With respect to the oceans, there are over 400 aquatic "dead zones" around the world, from the Baltic Sea to the Gulf of Mexico, from Lake Erie to the coastal waters of New Zealand. A dead zone is a region of water in which oxygen levels are too low for complex marine life to survive. They arise when human-created pollution from water treatment plants, runoff from farmland, and air pollution contaminate bodies of water, causing "blooms" of algae to grow.[11] Because individual alga live for only a short time, the corpses of the dead quickly accumulate. The problem is that the decay process absorbs the dissolved oxygen that fish and other marine creatures need to "breathe," and this results in suffocation. As of this writing, the biggest known dead zone is about 27,000 square miles.

This is a huge area, of course, but it's far smaller than the estimated size of the biggest *island of trash* in the ocean, the "Great Pacific Garbage Patch," located in the middle of the Pacific Ocean (south of Hawaii). Consisting mostly of small pieces of plastic "about the size of your pinkie fingernail,"[12] this trash heap is said to be somewhere between the size of Texas and "twice the size of the continental United States."[13] Countless organisms are sickened by it, as plastic bags and water bottles break down into microscopic inorganic particles via photodegradation. These particles are inadvertently consumed by small ocean critters, at which point they enter the food chain, eventually ending up in much larger organisms. The process of plastic photodegradation can also produce nasty toxins like PCBs and BPA, which can cause cancer, liver damage, and a wide range of additional health effects. Many other oceans around the world contain islands of trash, such as the Indian Ocean Garbage Patch and the North Atlantic Garbage Patch.

As for coral reefs, our best estimates today suggest that 60% are in danger of becoming underwater ghost towns; 10% are already gone as a result of overfishing and ocean acidification, an important consequence of increasing the amount of carbon dioxide in the atmosphere (see the next chapter). In fact, recent studies have shown that the acidification of the Pacific Ocean is literally causing the shells of "tiny marine snails that live along North America's western coast" to dissolve in the water.[14] This is an ominous sign, especially after a 2015 paper published in the journal *Science* found that the biggest mass extinction event in Earth's history, the Permian-Triassic Boundary extinction (sometimes referred to as the "Great Dying"), was largely caused by ocean acidification, as a result of increased levels of carbon dioxide in the atmosphere. Yet another study in *Science* extrapolated from current trends to the conclusion that by 2048, there may be no more wild-caught seafood in the ocean *at all*.[15] (The reader is welcome to extrapolate from the trends mentioned above, such as those pertaining to the global decline of vertebrates.)

Zooming out to the level of the ecosystem, a 2012 paper titled "Approaching a State Shift in Earth's Biosphere," authored by over twenty scientists and published in *Nature*, argues that we may be on the verge of a catastrophic, irreversible collapse of the global ecosystem. There are, the paper observes, two types of ecosystem collapse, or "state shifts." One involves the sudden destruction of an area from an exogenous force, such as when a bulldozer barrels through a forest and decimates everything in the area. This is called the *sledgehammer effect*. The second is far more insidious: it occurs when small changes, none of which have any great effect on the environment themselves, add up over time. This can lead the ecosystem across a critical threshold, or *tipping point*, after which sudden and dramatic changes occur, resulting in a state of total ruination.

These state shifts are well-understood in local ecosystems. What the paper does is apply this idea to the global ecosystem as a whole, and indeed they argue that there are empirical reasons for thinking the second kind of shift may be happening right now on the global scale, as a result of "human population growth with attendant resource consumption, habitat transformation and fragmentation, energy production and consumption, and climate change." The paper concludes that "comparison of the present extent of planetary change with that characterizing past global-scale state shifts, and the enormous global forcings [i.e., influence] we continue to exert, suggests that another global-scale state shift is highly plausible within decades to centuries, if it has not already been initiated." As a result, "the biological resources we take for granted at present may be subject to rapid and unpredictable transformations within a few human generations."[16] The world in which our children—or grandchildren—live may thus be radically different from the world that humanity has occupied since the Holocene began, at which point the last ice age ended and we entered a 12,000-year period of unusually comfortable weather (see chapter 11).

Not on My Planet (NOMPism)

Our species has brought about profound changes in the environment over the course of only a few centuries—*much too fast for evolution to keep up*. In the blink of an eye, from an evolutionary point of view, we've leveled forests, polluted the oceans, fragmented ecosystems, overfished the oceans, degraded a quarter of the world's land,[17] altered the chemical makeup of the atmosphere, increased the average surface temperature of the planet, melted glaciers, and initiated the sixth mass extinction event. If the global ecosystem were to cross a threshold, we might find ourselves as disoriented as an Inuit suddenly placed in the middle of the Arabian Desert, or a Bedouin relocated to northern Siberia. Survival would be difficult, and even if our species managed to persist, there would be significant compromises to our way of life.

Biodiversity matters because our civilization depends upon it. Just as we are minds *embodied*, so too are we bodies *environed*—meaning that our fate is inextricably bound up with the environment's. As a 2005 Millennium Ecosystem Assessment report explores in detail, consequences of biodiversity loss include food insecurity, "an increase in human suffering and economic losses from natural disasters," less balanced diets and therefore lower health scores, greater risk of infectious disease, energy insecurity (especially in developing countries), reduced quality and availability of water, social unrest, job losses, and even less ecotourism, which is among the "fastest-growing segments of tourism worldwide" and is very important for some rural communities. Biodiversity even "contributes to a range of other industries, including pharmaceuticals, cosmetics, and horticulture."[18]

The trimming of the biosphere could, furthermore, inflate the probability that other existential risks explored in this book will occur. For example, two countries with restricted access to food supplies could become highly motivated to develop nanoweaponry, in an effort to gain a strategic advantage. As described in chapter 4, weaponized nanotech is not like the nuclear threat, since it can have more targeted effects and can potentially facilitate a quick recovery after battle. These lowered barriers combined with an increasingly desperate ecological plight might be enough to push civilization across the threshold of self-destruction.

We should, at the very least, expect environmental degradation to cause the collapse of one or more societies in the coming century. As a recent NASA-funded article in *Ecological Economics* argues, the level of current resource overconsumption in the United States—along with socioeconomic stratification, an issue we'll return to later—is *sufficient* to cause the eventual implosion of our empire. Historically speaking, this has happened many times before, and "advanced, sophisticated, complex, and creative civilizations" are no exception to the rule: like bacteria in a petri dish, humans consume all the available resources until the environment can no longer support them, and society crumbles. The paper further argues that we shouldn't expect future technologies to save us, since while "technological change can raise the efficiency of resource use . . . it also tends to raise both per capita resource consumption and the scale of resource extraction."[19] Invention is often the mother of necessity, not the other way around.

What's frustrating about the risks posed by biodiversity loss is that (i) we have a good understanding of what's driving it and what its consequences for civilization are likely to be, and (ii) there are meaningful actions we can take to significantly reduce current biodiversity loss trends. Yet environmental regulations, green technology, and sustainable living continue to be scoffed at by a large portion of the American public, which values short-term economic growth over the livability of our planet in a few generations.

Notes

1.   Some of this is taken *ad verbum* from "AGU 2012: American Geophysical Union Fall Meeting," Inside GNSS, http://www.insidegnss.com/node/3297.

2.   See Brian Merchant, "Scientist: It's a Statistical Probability That Earth Is Fucked," *Motherboard*, December 10, 2012, http://motherboard.vice.com/blog/scientist-its-a-statistical-probability-that-earth-is-fucked.

3.   See Kevin Plaxco and Michael Gross, *Astrobiology: A Brief Introduction* (Johns Hopkins University Press, 2011), p. 188.

4.   See Lee Sweetlove, "Number of Species on Earth Tagged at 8.7 Million," *Nature* (2011), doi:10.1038/news.2011.498.

5.   See "The Extinction Crisis," Center for Biological Diversity, http://www.biologicaldiversity.org/programs/biodiversity/elements_of_biodiversity/extinction_crisis/.

6.   See "Article 2: Use of Terms," Convention on Biological Diversity, accessed June 18, 2015, https://www.cbd.int/convention/articles/default.shtml?a=cbd-02.

7.   See the "Global Biodiversity Outlook 3," Convention on Biological Diversity, 2010, https://www.cbd.int/doc/publications/gbo/gbo3-final-en.pdf.

8.   See "Living Planet Report 2014," World Wildlife Fund, October 2014, http://wwf.panda.org/about_our_earth/all_publications/living_planet_report/.

9.   See, respectively, "Half of World's Primate Species Endangered, Report Says," *CNN*, Feburary 18, 2010, http://edition.cnn.com/2010/TECH/science/02/17/endangered.species/index.html?_s=PM:TECH; "The Extinction Crisis," Center for Biological Diversity, http://www.biologicaldiversity.org/programs/biodiversity/elements_of_biodiversity/extinction_crisis/; and John Vidal, "One in Five Reptile Species Faces Extinction—Study," *Guardian*, February 15, 2013, http://www.theguardian.com/environment/2013/feb/15/reptile-species-face-extinction.

10.   See "Global Biodiversity Outlook 3," Convention on Biological Diversity, 2010, https://www.cbd.int/doc/publications/gbo/gbo3-final-en.pdf, p. 24.

11.   To be clear, a dead zone could occur naturally, but it's more likely to be anthropogenic.

12.   See Annalee Newitz, "Lies You've Been Told about the Pacific Garbage Patch," *io9*, May 5, 2012, http://io9.com/5911969/lies-youve-been-told-about-the-pacific-garbage-patch.

13.   See Kathy Marks and Daniel Howden, "The World's Rubbish Dump: A Tip That Stretches from Hawaii to Japan," *Independent*, February 5, 2008, http://www.independent.co.uk/environment/green-living/the-worlds-rubbish-dump-a-tip-that-stretches-from-hawaii-to-japan-778016.html.

14.   See Eli Kintisch, "Snails Are Dissolving in Pacific Ocean," *Science*, May 1, 2014, http://news.sciencemag.org/biology/2014/05/snails-are-dissolving-pacific-ocean.

15.   See Juliet Eilperin, "World's Fish Supply Running Out, Researchers Warn," *Washington Post*, November 3, 2006, http://www.washingtonpost.com/wp-dyn/content/article/2006/11/02/AR2006110200913.html.

16.   See Anthony Barnosky, et al. "Approaching a State Shift in Earth's Biosphere," *Nature* 486, no. 52–58 (2012), http://www.nature.com/nature/journal/v486/n7401/full/nature11018.html.

17.   See the "Global Biodiversity Outlook 3," Convention on Biological Diversity, 2010, https://www.cbd.int/doc/publications/gbo/gbo3-final-en.pdf.

18.   See Millennium Ecosystem Assessment, *Biodiversity & Human Well-being: Biodiversity Synthesis* (Washington, DC: World Resources Institute, 2005), pp. 31–32, http://www.millenniumassessment.org/documents/document.354.aspx.pdf.

19.   See Nafeez Ahmed, "NASA-funded Study: Industrial Civilisation Headed for 'Irreversible Collapse'?" *Guardian*, March 14, 2014, http://www.theguardian.com/environment/earth-insight/2014/mar/14/nasa-civilisation-irreversible-collapse-study-scientists.

8

Warming Up to Extinction

• • • • •

Global Warming Is the Climate's Change

In chapter 2, we discussed the Doomsday Clock, which a group of physicists created after World War II to symbolize our nearness to a global catastrophe. While the clock was intended to reflect the probability of our species' suicide through nuclear annihilation, those responsible for setting the minute hand have recently included another possibility. This newly added concern is *global warming*, a phenomenon whose reality is, to be clear, still hotly debated—although not among the experts in the fields of climatology, biology, and geology. As the *Bulletin of the Atomic Scientists'* board of directors announced in 2007, when the Doomsday Clock's minute hand was moved from 7 to 5 minutes before midnight, "As in past deliberations, we have examined other human-made threats to civilization. We have concluded that the dangers posed by climate change are nearly as dire as those posed by nuclear weapons."[1] In early 2015, the *Bulletin* moved the minute hand even further toward doom, from 5 to 3 minutes. The very first reason given for this action was, in fact, "unchecked climate change."[2]

Global warming is intimately tied to the risks of the previous chapter. Along with pollution, invasive species, overfishing, the fragmentation of habitats, and sea-level rise, it's one of the major causes of biodiversity loss. The term "global warming" refers to the average increase in surface temperatures observed around the world. The word "average" here is important, because many people draw erroneous conclusions about global trends in the *climate* from local trends in the *weather*. The fatal flaw is that local and global trends don't always line up: it's entirely possible for one region of the planet, say, Europe, to actually get *colder* as the globe as a whole starts to cook. By analogy, the average grade of a class consisting of

20 students could rise even as the grades of several students fall. There is no contradiction here!

Indeed, one of the most intense winters along the East Coast of the US occurred in 2009 and 2010. At the beginning of February, a Category 3 nor'easter dumped record-breaking amounts of snow from Virginia to New York. The media dubbed this meteorological monstrosity "Snowmageddon." It prompted a large number of climate change denialists—truly dangerous people whose intransigent resistance to science threatens the livability of our planet—to argue that global warming isn't really happening. After all, they claimed, if it were happening, how could there be so much snow on the ground? Sean Hannity of Fox News declared that Snowmageddon "would seem to contradict Al Gore's hysterical global warming theories."[3] (Note that Gore didn't invent the theory of global warming, so Hannity is doubly wrong here.) More recently, in 2015, the Republican senator Jim Inhofe walked into the Senate and threw a snowball, made from snow outside, on the floor "as a visual rebuttal to claims of global warming."[4]

But 2010—the year of Snowmageddon—turned out to be the *hottest year ever* on record (tying with one other recent year). In other words, while the East Coast was digging out from an oppressive amount of snow, the globe was sizzling more than it has in quite possibly the past 2 million years. This heat record was later broken in 2014, during which "the average temperature across global land and ocean surfaces was 1.24°F (0.69°C) above the 20th century average."[5] To get a sense of what this means, the difference between our present climate and a full-on ice age is a mere 7 degrees Fahrenheit.[6] So a 1.24-degree rise in such a short time is a *big deal*. With only a single exception, the top eleven hottest years on record have all occurred in the twenty-first century. Behind 2014 and 2010, the next hottest years are: 2005, 1998, 2013, 2003, 2002, 2006, 2009, 2007, and 2004.[7] As I write this, 2015 is on track to beating the record set by 2014.

Global warming is happening: this is a known fact. But what's causing it? There is, of course, natural variability in Earth's climate over geological timescales. *No one knows this better than climatologists themselves.* Rumors of past climate change have, in fact, led some nonexperts to assert that *even if* global warming is occurring, it's the result of nonanthropogenic forces. For example, the almost-VP and former governor of Alaska, Sarah Palin, tweeted in 2009: "Earth saw clmate chnge4 ions;will cont 2 c chnges.R duty2responsbly devlop resorces4humankind/not pollute&destroy;but cant alter naturl chng." Aside from the embarrassment of Palin not knowing the difference between "ions" and "eons," the best current evidence unequivocally shows that human activity is almost entirely responsible for the observed climatic changes. Indeed, not only are we causing the climate to warm up, we're causing it to warm up *very quickly* relative to evolutionary timescales. This is catastrophic because, as discussed earlier, natural selection works *very slowly*, and consequently it can't keep up with rapid environmental shifts. The result is a loss of population size and, at the extreme, the permanent extinction of species.

The main driver of global warming is a molecule called *carbon dioxide* (CO_2). As is commonly known, CO_2 is a byproduct of burning fossil fuels. What happens is this: the sun releases electromagnetic radiation that travels some 93 million miles in about 8 minutes to Earth. Most of this radiation takes the form of *visible light*, which is what our eyes evolved to detect.[8] The earth absorbs a certain amount of this light, and then reradiates it back into the atmosphere. The darker the surface, the more visible light is absorbed. In the case of snow, hardly any visible light is absorbed, as white is nothing but all the visible frequencies of radiation, or colors, added together. The catch is that the light reradiated by the earth isn't visible light: it's *infrared light*, composed of frequencies we can only detect through the sense of touch, as heat.

CO_2 enters the picture here because *it's transparent to visible light but not to infrared light*. This means that if you have lots of CO_2 in the atmosphere, visible light from the sun will pass right through it, but the infrared light reradiated from the earth *won't* make it back out into space. The result is heat being trapped inside the atmosphere: global warming. This is called the "greenhouse effect" because glass panels do the same thing as CO_2: they let visible light through, but trap infrared light trying to get back out.[9] Gases that cause this effect are called "greenhouse gases" (GHGs).

As of this writing, China is the world's biggest producer of greenhouse gases, with the United States in second place. In terms of per capita carbon output, Qatar is the worst offender, with the United States in eighth.[10] Furthermore, just as wealthy countries have a significantly greater environmental impact than poor ones, wealthy individuals—from Bill Gates and Kanye West to Donald Trump and even Jon Stewart—are more responsible for the rise in atmospheric carbon dioxide than those with lower incomes. The fact that poor people will almost certainly suffer the consequences of global warming more than rich folks is at the heart of the concept of *climate justice*, a version of social justice.[11]

The impact of climate change on the lives of our children will be profound. Negative consequences include more intense heat waves, longer and more damaging wildfire seasons, heavier rainstorms and flooding, the spread of insect-borne diseases, the melting of glaciers and the polar ice caps, sea-level rise, deforestation, desertification, significant food-supply disruptions, higher food prices, more extreme weather, including "as many as 20 more hurricanes and tropical storms each year by the end of the century,"[12] significant coastal flooding, and megadroughts lasting decades,[13] and an overall increase in air pollution.[14] Scientists have even stated that global warming will lead to "longer and more intense allergy seasons"[15] and a 50% greater chance of getting hit by lightning at the end of the century.[16] According to the most recent and comprehensive summary of climate change research, global warming will be not only "severe" and "pervasive" but "irreversible."[17] Another paper, published in 2009, reports that "climate change that takes place due to increases in carbon dioxide concentration is largely irreversible for 1,000 years after emissions stop."[18] I hope history won't forget those who've chosen to ignore the evidence and keep the status quo of fossil fuel consumption in place.

Positive Feedback Is Unwelcome

The most worrisome global warming–related possibility is that a *runaway greenhouse effect* turns our Earth into Venus—a planet that's similar to Earth in size and chemical constitution but which has up to 870-degree surfaces due to out-of-control global warming. This could result if one or more self-amplifying positive feedback loops were to become activated. In the case of Venus, as the planet's surface became hotter, more water evaporated into the air. Since water vapor is a greenhouse gas, it allowed visible light through but trapped the reradiated infrared light. This made the planet's surface even hotter, which caused more water to evaporate—and so on, until Venus became the hellish cauldron it is today.

On Earth, the worry isn't so much water vapor but methane. We know, for example, that huge quantities of methane are entombed in permafrost, and that this frozen soil in the northern regions of the globe is beginning to thaw. If these subterranean deposits were to be released into the atmosphere, they would exacerbate climate change, as methane is an even more potent greenhouse gas than carbon dioxide.[19] (So is water vapor, incidentally.) The result would be more permafrost thawing and, consequently, more methane being released into the air. The globe could warm up at an exponential pace, as if one's thermostat was set to *increase* the temperature even more as the house gets *hotter*.

There are also large amounts of methane trapped under the deep blue ocean. In 2014, a group of researchers in the Arctic discovered, much to their surprise, "vast methane plumes escaping from the seafloor." Some of these "bubbles" were even seen reaching the ocean's surface.[20] This is very unsettling: it implies that global warming may be a ticking time bomb just waiting to explode. As the glaciologist Jason Box commented on this discovery—expressing a view similar to Werner's—"If even a small fraction of Arctic sea floor carbon is released to the atmosphere, we're f'd."

The best estimates suggest that the probability of a runaway greenhouse effect happening on Earth is not particularly high, but research is ongoing.[21] We at least know it's physically *possible*, since it happened on Venus. Perhaps melting permafrost and Arctic methane plumes will make us look closer at this possibility.

In sum, climatologists see global warming as a clear and present danger. It is one of the major causes of biodiversity loss, and therefore of the Holocene extinction event. While we might still be able to avert a global catastrophe, or worse, there's a growing chorus of scientists who say we've already passed the point of no return.

Notes

1.   See Tom Zeller Jr., "The End Is Nearer, Say Atomic Scientists," *New York Times*, January 17, 2007, http://thelede.blogs.nytimes.com/2007/01/17/the-end-is-nearer-say-atomic-scientists/comment-page-3/.

2.   See "Press Release: It Is Now 3 Minutes to Midnight," *Bulletin of the Atomic Scientists*, January 22, 2015, http://thebulletin.org/press-release/press-release-it-now-3-minutes-midnight7950.

3.   See Daniel Schulman, "Snowpocalypse: Take That Al Gore!," *Mother Jones*, February 10, 2010, http://www.motherjones.com/mojo/2010/02/gop-snowpocalypse-global-warming-al-gore.

4.   Quoted from Jack Kenny, "Inhofe Throws Snowball in Senate to Refute Global Warming," *New American*, March 2, 2015, http://www.thenewamerican.com/usnews/congress/item/20237-inhofe-throws-snowball-in-senate-to-refute-global-warming.

5. See "Summary Information," National Centers for Environmental Information, http://www.ncdc.noaa.gov/sotc/summary-info/global/2014/12.

6. If "sustained over a long period" of time. See "Super-eruptions, Global Effects and Future Threats," Geological Society of London Working Group, http://www.geo.mtu.edu/~raman/VBigIdeas/Supereruptions_files/Super-eruptionsGeolSocLon.pdf.

7. See Andrea Thompson, "2014 on Track to Be Hottest Year on Record," *Scientific American*, September 23, 2014, http://www.scientificamerican.com/article/2014-on-track-to-be-hottest-year-on-record/.

8. This is why our eyes evolved to see visible light: it makes up most of the light from the sun.

9. There's a bit more to this, since greenhouses also prevent convection from occurring. But the point stands.

10. China and the United States just reached an "historic" agreement on CO2 emissions, but even this agreement does not appear to be enough to meet the goal of the 2009 Copenhagen summit: prevent global temperatures from rising more than 3.6 degrees. (See Edward Wong, "China's Climate Change Plan Raises Questions," *New York Times*, November 12, 2014, http://www.nytimes.com/2014/11/13/world/asia/climate-change-china-xi-jinping-obama-apec.html?_r=0.)

11. It's also worth noting that the study mentioned in the previous chapter about ocean acidification and the Permian-Triassic Boundary extinction found that the rise in atmospheric CO2 at the time (which caused the ocean to become more acidic) occurred "at a rate similar to modern emissions." See "Greatest Mass Extinction Drive by Acidic Oceans, Study Finds," *Science Daily*, April 9, 2015, http://www.sciencedaily.com/releases/2015/04/150409143033.htm.

12. See "11 Ways Climate Change Affects the World," *CNN*, April 22, 2015, http://www.cnn.com/2014/09/24/world/gallery/climate-change-impact/.

13. See Darryl Fears, "A 'Megadrought' Will Grip US in the Coming Decades, NASA Researchers Say," *Washington Post*, February 12, 2015, http://www.washingtonpost.com/national/health-science/todays-drought-in-the-west-is-nothing-compared-to-what-may-be-coming/2015/02/12/0041646a-b2d9-11e4-854b-a38d13486ba1_story.html.

14. Many of these consequences of global warming are recapitulated here *ad verbum* from "Global Warming Impacts: The Consequences of Climate Change Are Already Here," Union of Concerned Scientists, ahttp://www.ucsusa.org/our-work/global-warming/science-and-impacts/global-warming-impacts#.VYHCzxNViko.

15. See Kim Krisberg, "Leaders Make Link between Climate Change Effects, Health of Americans: New Initiatives," *Nation's Health* 45, no. 5 (2015): pp. 1–14.

16. See David Romps et al., "Projected Increase in Lightning Strikes in the United States Due to Global Warming," *Science* 346, no. 6211 (2014): p. 851.

17. See Justin Gillis, "UN Panel Issues Its Starkest Warning Yet on Global Warming," *New York Times*, November 2, 2014, http://www.nytimes.com/2014/11/03/world/europe/global-warming-un-intergovernmental-panel-on-climate-change.html?_r=0.

18. See Suan Solomon et al., "Irreversible Climate Change Due to Carbon Dioxide Emissions," *PNAS* (2008), doi: 10.1073/pnas.0812721106.

19. It's about 30 times more potent, in fact.

20. See SWERUS-C3, "SWERUS-C3: First Observations of Methane Release from Arctic Ocean Hydrates," http://www.swerus-c3.geo.su.se/index.php/swerus-c3-in-the-media/news/177-swerus-c3-first-observations-of-methane-release-from-arctic-ocean-hydrates.

21. See, for example, Nafeez Ahmed, "James Hansen: Fossil Fuel Addition Could Trigger Runaway Global Warming," *Guardian*, July 10, 2013, http://www.theguardian.com/environment/earth-insight/2013/jul/10/james-hansen-fossil-fuels-runaway-global-warming.

9

Caldera and Comets

· · · · ·

The bright sun was extinguish'd and the stars
Did wander darkling in the eternal space,
Rayless, and pathless, and the icy earth
Swung blind and blackening in the moonless air

—*Lord Byron*[1]

A Meteorological Frankenstein

In late spring of 1816, Mary Godwin and her lover Percy Bysshe Shelley were visiting their friend Lord Byron near the crystal waters of Lake Geneva, Switzerland.[2] Their plan was, most likely, to spend the summer hiking around the Swiss Alps, relaxing amid the snowcapped mountains and flower-filled meadows that pop up that time of year. Much to their surprise, though, the summer of 1816 was marked by incessant rain, cold temperatures, and dark, cloudy skies. As Mary wrote in a letter to her sister, "One night we enjoyed a finer storm than I had ever before beheld. The lake was lit up—the pines on Jura made visible and all the scene illuminated for an instant, when a pitchy blackness succeeded, and the thunder came in frightful bursts over our heads amid the darkness."[3] The unusually gloomy weather kept the group cooped up indoors for months, reading each other ghost stories to pass the time. One day, Lord Byron proposed a challenge to his visitors: try to write a ghost story of their own to share. With nothing else to do, Mary Godwin began work on a novel—originally intended to be a short story—that would later become a canonical masterpiece of literary fiction: *Frankenstein*.[4]

Switzerland wasn't the only place experiencing odd weather. On June 4, 1816, residents of Connecticut and New Jersey woke up to a frost. Two days later it snowed in Albany, New York. In July and August, ice formed on rivers and lakes as far south as Pennsylvania. People across the northeastern US reported a lingering "dry fog" in the air that wouldn't lift even with wind and rain.[5] The sky was transformed into an eerily glowing crimson dome. These strange phenomena—and the resulting loss of food and livestock—led many in New England to migrate south and west. It was this migration, in 1816, that was partly responsible for the American Heartland becoming populated. (In fact, it led the young Joseph Smith—founder of the Mormon religion—to move west with his family from Vermont to New York, where he received his first vision at the age of 14.)

Back in Europe, frigid air and heavy precipitation brought about widespread food shortages. Many people, cold and hungry, rioted, with French peasants ambushing "grain carts on the way to markets."[6] In Ireland, rain fell for eight weeks straight, causing famine and malnutrition.[7] This led to a severe outbreak of typhus, which swept across the country, killing over 100,000 people. Further east, an epidemic of cholera devastated India, eventually spreading "north to Europe and west to Egypt."[8] China and other areas of Asia suffered large crop failures, and many people starved as a result. The situation was so dire in the province of Yunnan that "Yunnanese resorted to eating white clay, while parents sold their children in the town markets or killed them out of mercy."[9] Around the globe, from East to West, violence and social upheaval dominated 1816, as food insecurity resulting from anomalous weather pushed whole societies into a morass of desperation.

Why was the weather of 1816 so freaky? What caused this "Year Without a Summer," as it's now called? The answer takes us far away in both space and time, to the other side of the hemisphere slightly more than a year before. It takes us to the beautiful island of Sumbawa, located in the Indonesian archipelago, where a protrusion of Earth called Mount Tambora had been rumbling for several years prior. On April 5, 1815, it began spewing ash into the sky. This produced a series of explosions so thunderous that soldiers literally hundreds of miles away heard them and became alarmed, thinking they might be the growls of war. A group stationed on the Indonesian island of Java even rushed to the aid of another post, thinking that it had been attacked.[10]

For days Mount Tambora gurgled, threatening a violent outburst of volcanic pyrotechnics. Then, on April 10, a plume of smoke climbed some 25 miles into the atmosphere. Three pillars of fire shot out of Tambora, eventually merging into a single column of blazing rock. Toxic ash and pumice eight-inches wide rained down upon Sumbawa, and a tsunami crashed into the beaches of nearby islands. Dead vegetation entangled with buoyant pumice created massive "rafts" floating on the ocean, some over three-miles wide.[11] Over 10,000 people on the island "died instantly" from the blast, while many more died in the aftermath, due to starvation and disease.[12] (In fact, Tambora means "gone" in the native language.) While the local culture was decimated, the worst effects were yet to come, as Tambora's fury would, more than a year later, lead to the global chill of "Eighteen Hundred and Froze to Death."

Supervolcanism

Mount Tambora wasn't an existential catastrophe for humanity, of course, but it does illustrate how a volcanic eruption on one side of the planet can bring about significant changes in the global climate, resulting in widespread famine, disease, social unrest, and political instability on the other. As it happens, Tambora wasn't the only volcano in Indonesia to have had a major impact in our species' 200,000-year history. About 74,000 years ago, according to one estimate, the Indonesian island of Sumatra produced a massive *supereruption*.[13] This event of geological hostility is called the Toba catastrophe, and it was so enormous that scientists actually coined the term "supereruption" to describe it.

The Toba catastrophe might have resulted in up to a decade of unusual weather, with average surface temperatures sinking by 5.4 to 9 degrees F.[14] This appears to have had a nontrivial effect on the flora and fauna at the time (although there is some recent evidence suggesting the effects weren't so severe[15]). The geologist Michael Rampino calculates that up to "three-quarters of the plant species in the Northern Hemisphere perished," and a number of studies have linked the Toba catastrophe with animal extinctions.[16] The eruption also appears to have coincided, historically, with a significant bottleneck in the population of *Homo sapiens*. Some research suggests that as few as 500 breeding females survived, and the total population of our species may have been "as small as 4000 for approximately 20,000 years."[17] Few people today realize just how close our species might have come to extinction in the Pleistocene.

When the Toba supereruption occurred, it emptied a chamber once occupied by magma. Once this molten rock was ejected, the chamber collapsed in on itself to form a *caldera*, which later filled with water, becoming present-day Lake Toba. The volcanic winter it produced was not unlike the nuclear winter phenomenon discussed in chapter 2. What happens is this: when a supervolcano explodes, it releases large quantities of sulfur dioxide into the stratosphere. These particles react with water droplets to become sulfuric acid, which behave like tiny *molecular mirrors* by reflecting incoming light from the sun. The result is that less solar energy reaches Earth, causing the earth's surface temperature to plummet for weeks, months, or years. In addition, sulfuric acid can also "absorb heat from the Earth, [thereby] warming the upper atmosphere and cooling the lower atmosphere," and it can contribute to the destruction of the ozone layer.[18]

Counting all the caldera on Earth reveals that supereruptions occur on average about once every 50,000 years.[19] This means it's relatively unlikely that one will happen in our lifetimes, but a low probability doesn't imply, much less guarantee, that we're safe. If a Toba-sized event were to occur tomorrow—perhaps in Yellowstone National Park, where a supervolcano has erupted 3 times over the past 2 million years—it could seriously disrupt the global economy, undermine the stability of countries, and cause wars to break out. Global crop failures lasting several years or more could foment riots and violent clashes between people desperate for food and other commodities. Countries could struggle to maintain law and order within their borders, as large numbers of people face starvation, especially in the developing parts of the world. Widespread malnutrition could simultaneously increase the probability of catastrophic outbreaks of infectious disease.[20] And dwindling resources could nudge countries to war, potentially leading to the realization of one or more scenarios discussed above and below, such as a targeted nanotech attack or a nuclear exchange between rival states.

Despite the relative infrequency of supereruptions, their effects on modern civilization could be significant. Recall from chapter 1 that a risk is the probability of an event multiplied by its *consequences*. Thus, a risk can be significant even if its probability is low. This is why some geologists have taken an activist stance and urged the world's governments to take the threat from supervolcanoes seriously. Given how much larger the world population is today than when Mary Shelley wrote *Frankenstein*, the consequences of even a Tambora-sized event could be quite catastrophic. (By comparison, Toba "ejected about 300 times more volcanic ash" than Tambora![21]) As Rampino avers, "volcanic super-eruptions pose a real threat to civilization, and efforts to predict and mitigate volcanic climatic disasters should be contemplated seriously."[22]

This sentiment is echoed in a 2005 publication by the Geological Society of London, which observes that while we're helpless to defend ourselves against a supereruption ("even science fiction cannot produce a credible mechanism for averting a super-eruption"), there are meaningful steps we can take to anticipate a major eruption and prepare for the aftermath. "Sooner or later a super-eruption will happen on Earth," the publication states, "and this is an issue that . . . demands serious attention."[23] A supervolcano very nearly wiped out our species once before; next time we might not be so lucky.[24]

The Impact of an Impactor

Whereas supervolcanoes are threats from the earth, asteroids and comets are hazards from the heavens. Shortly after forming some 4.5 billion years ago, the earth narrowly escaped total destruction: a giant celestial body the size of Mars (about half the diameter of Earth) hurdled through space and collided with our young, still moonless planet. The impact was so forceful that the atmosphere was likely blown off by the collision.[25] Billions of rocky fragments were blasted into space, some of which turned around (due to gravity) and reimpacted Earth. Other bits of planetary shrapnel, though, coalesced into a ball of magma that neither fell back to Earth nor wandered off into the solar system. Over time, this ball cooled to become a dusty sphere, and today it orbits the earth about once every month (or *moonth*). It was through this act of astronomical violence that we got our moon.[26]

There's no known risk of Earth encountering an impactor the size of Mars (huge!) in the foreseeable future. But we wouldn't need to collide with such an object to suffer catastrophic consequences. In 1908, for example, an asteroid only 200-feet wide exploded over Siberia with "the force of a hydrogen bomb." If this had happened above a city, millions could have died.[27] More recently, an object from space traveling at about 42,000 mph shot through the atmosphere above the Russian city of Chelyabinsk, giving off more light than the sun before it burned up. A number of dashcam videos, many of which have been uploaded to YouTube, managed to record this brilliant—and terrifying—display of asteroidal fireworks.

It's also worth mentioning a near miss in 1972 when a meteoroid essentially bounced off Earth's atmosphere over the western United States (sort of like a stone skipping across a lake), passing within 35 miles of the ground over Alberta, Canada.[28] If this "Great Daylight Fireball" had struck our planet, the resulting explosion could have been comparable to a small nuclear bomb. And given the year—in the middle of the Cold War—an impact like this could have initiated a quick nuclear retaliation from the US, having misinterpreted a projectile from space for a missile from Russia.[29] (Even today, such an event could conceivably initiate a nuclear exchange, if cooler heads don't prevail.) While this potential impactor was the size of a small truck, the comet that smashed into the Yucatán Peninsula some 65 million years ago was roughly 6 miles in diameter.[30] It barreled all the way into Earth's mantle with the force of 100 million megatons of TNT and, as discussed in chapter 7, was responsible for the extinction of nearly all the dinosaurs.

In the last half a billion years, Earth has encountered seven or more relatively large rogue objects from space, each of which was "sufficiently severe to result in a wave of extinction."[31] As of this writing, there are exactly 1,574 *potentially hazardous asteroids* (PHAs) circling like vultures within the solar system. (This number is updated daily on the Minor Planet Center's website.) These are a particularly worrisome variety of near-Earth objects (NEOs), since they could, by definition, desolate a sizable region of the world, wiping out whole cities or coastlines.[32] In the worst-case scenario, an impactor of sufficient mass could collide with Earth, propel large amounts of debris into the atmosphere, and initiate an extended period of cold called an *impact winter*. As with supereruptions, the main consequences would be agricultural failures, famine, infectious disease, sociopolitical instability, economic collapse, and, in a phrase, "severe damage to the underpinnings of civilization."[33] If big enough, an impactor could even lead to the extinction of our species.

How likely is such a doomsday event? What's the probability of an assassin from space flooding the sky with ash and dust or landing in the ocean to produce a massive tsunami? The astrophysicist Neil deGrasse Tyson writes in a *Wired* article that "the chances that your tombstone will read 'Killed by an Asteroid' are about the same as they'd be for 'Killed in an Airplane Crash.'"[34] According to one study, an impactor capable of taking out "more than 1.5 billion people through direct and non-direct effects" occurs about once every 100,000 years, roughly half as often as supereruptions.[35] Despite their rarity, impactors (as well as supervolcanoes) can be tied to many extinction events in history, including a few of the "Big Five." In 2013, NASA claimed to be spying on roughly 95% of "the largest asteroids that could endanger Earth—space rocks at least 0.6 miles (1 km) wide."[36] This means that there are still some big surprises that could leap out from the darkness and damage civilization.[37] As the current administrator of NASA, Charles Bolden, puts it, if a large asteroid were about to crash into the United States in three weeks, the only thing to do is "pray."[38]

Interestingly, there are technologies we could develop to either destroy or deflect an incoming impactor. But as Alan Hale points out, these technologies are *dual-use*: if they can deflect objects away from Earth, then they can deflect them toward Earth as well.[39] The danger here is presently low, but it could nontrivially rise, speaking speculatively, if future nanofactories open up the space frontier (as discussed in chapter 4), thereby making it possible for terrorists or even single individuals to manufacture spacecraft. Some spacecraft could thus be designed to redirect an asteroid toward a particular city or state, or even to collide with a space station orbiting Earth.

In sum, both supervolcanoes and asteroids/comets present ongoing risks to human survival. Although neither is especially probable, their potential consequences are significant enough to warrant genuine concern. Unlike many other risks in this book, there are fairly clear actions that can be taken to either avoid or mitigate the effects of such risks. Yet few of these actions are being taken. In some cases, the actions are worrisomely inadequate. For example, an internal watchdog in 2014 slammed NASA's Near Earth Objects project, saying that (to quote one article) "the program is nowhere near meeting its stated goal" of cataloguing 90% of all near-Earth objects by 2020.[40] As Tyson grimly notes in the same *Wired* article, we humans would be the laughingstock of the universe if, despite our big brains and advanced space program, we met the same fate as the dinosaurs.[41]

Notes

1.   This comes from Lord Byron's "Darkness," written in June 1816. I am indebted to the Geological Society of London for pointing this out in a publication. I also found, later on, that this poem opens Clive Oppenheimer, "Climatic, Environmental and Human Consequences of the Largest Known Historic Eruption: Tambora Volcano (Indonesia) 1815," *Progress in Physical Geography* 27, no. 2 (2003).

2.   Mary Godwin married Percy later that year, but she was already calling herself "Mrs. Shelley" on the Geneva trip.

3. See Nell Greenfieldboyce, "Did Climate Inspire the Birth of a Monster?" *NPR*, August 13, 2007, http://www.npr.org/templates/story/story. php?storyId=12688403.

4. See John Clubbe, "The Tempest-toss'd Summer of 1816: Mary Shelley's *Frankenstein*," *Byron Journal*, 19 (1991), 26–40, http://knarf.english.upenn.edu/ Articles/clubbe.html; and Nell Greenfieldboyce, "Did Climate Inspire the Birth of a Monster?" *NPR*, August 13, 2007, http://www.npr.org/templates/story/story. php?storyId=12688403.

5. See Clive Oppenheimer, "Climatic, Environmental and Human Consequences of the Largest Known Historic Eruption: Tambora Volcano (Indonesia) 1815," *Progress in Physical Geography* 27, no. 2 (2003): p. 244.

6. See Dennis Pirages, "Nature, Disease, and Globalization: An Evolutionary Perspective," in *Globalization as Evolutionary Process: Modeling Global Change*, ed. George Modelski, Tessaleno Devezas, and William Thompson (Routledge, 2008), p. 230.

7. See Robert Evans, "Blast from the Past," *Smithsonian Magazine*, July 2002, http://www.smithsonianmag.com/history/blast-from-the-past-65102374/?no-ist= &page=1.

8. Again, see Dennis Pirages, "Nature, Disease, and Globalization: An Evolutionary Perspective," in *Globalization as Evolutionary Process: Modeling Global Change*, ed. George Modelski, Tessaleno Devezas, and William Thompson (Routledge, 2008), p. 230.

9. See Gillen D'Arcy Wood, "The Volcano That Changed the Course of History," *Slate*, April 9, 2014, http://www.slate.com/articles/health_and_science/ science/2014/04/tambora_eruption_caused_the_year_without_a_summer_ cholera_opium_famine_and.html.

10. See Robert Evans, "Blast from the Past," *Smithsonian Magazine*, July 2002, http://www.smithsonianmag.com/history/blast-from-the-past-65102374/?no- ist=&page=1. Also, from Sir Stamford Raffles' memoir.

11. See Richard Smothers, "The Great Tambora Eruption in 1815 and Its Aftermath," *Science* 224, no. 4654 (1984), doi:10.1126/science.224.4654.1191.

12. See Robert Evans, "Blast from the Past," *Smithsonian Magazine*, July 2002, http://www.smithsonianmag.com/history/blast-from-the-past-65102374/?no- ist=&page=1.

13. Russell Blackford puts the date at 70,000 in the foreword of this book. This is no less correct than the date I use here, since the best current estimates range from 69,000 to 77,000 years ago.

14. See Anne Casselman, "Ancient Humans in Asia Survived Super-Eruption, Find Suggests," *National Geographic News*, July 5, 2007, http://news.nationalgeographic.com/news/2007/07/070705-india-volcano.html.

15. See Charles Choi, "Supervolcano Not to Blame for Humanity's Near-Extinction," *Live Science*, April 29, 2013, http://www.livescience.com/29130-toba-supervolcano-effects.html.

16. See Robert Roy Britt, "Super Volcano Will Challenge Civilization, Geologists Warn," *Live Science*, March 8, 2005, http://www.livescience.com/200-super-volcano-challenge-civilization-geologists-warn.html.

17. See Michael Rampino, "Super-volcanism and Other Geophysical Processes of Catastrophic Import," in *Global Catastrophic Risks*, ed. Nick Bostrom and Milan Ćirković (Oxford University Press, 2008), p. 212. But, as Stephen Sparks pointed out to me in personal communication, genetic dating for this bottleneck is not particularly accurate, and could be off by a considerable amount.

18. See Steve Connor, "Yellowstone's Slumbering Giant," *Rense*, http://www.rense.com/general63/yellowstonesslumbering.htm.

19. See Michael Rampino, "Super-volcanism and Other Geophysical Processes of Catastrophic Import," in *Global Catastrophic Risks*, ed. Nick Bostrom and Milan Ćirković (Oxford University Press, 2008), p. 215.

20. For more, see Michael Rampino, "Super-volcanism and Other Geophysical Processes of Catastrophic Import," in *Global Catastrophic Risks*, ed. Nick Bostrom and Milan Ćirković (Oxford University Press, 2008), p. 213–214, which outline these possible consequences in even more detail.

21. See "Super-eruptions: Global Effects and Future Threats," Geological Society of London Working Group, p. 6.

22. See Michael Rampino, "Super-volcanism and Other Geophysical Processes of Catastrophic Import," in *Global Catastrophic Risks*, ed. Nick Bostrom and Milan Ćirković (Oxford University Press, 2008), p. 216.

23. See "Super-eruptions: Global Effects and Future Threats," Geological Society of London Working Group, p. 2.

24. This section relied heavily on Michael Rampino, "Super-volcanism and Other Geophysical Processes of Catastrophic Import," in *Global Catastrophic Risks*, ed. Nick Bostrom and Milan Ćirković (Oxford University Press, 2008), and "Super-eruptions: Global Effects and Future Threats," Geological Society of London Working Group.

25.   See Katia Moskvitch, "Giant Impact That Formed the Moon Blew Off Earth's Atmosphere," *Space.com*, October 1, 2013, http://www.space.com/23031-moon-origin-impact-earth-atmosphere.html.

26.   This is currently the accepted theory. But there remains some doubt about whether it's correct; to date, no plausible alternative hypotheses have been proposed.

27.   See Richard Posner, "Public Policy Towards Catastrophe," in *Global Catastrophic Risks*, ed. Nick Bostrom and Milan Ćirković (Oxford University Press, 2008), p. 185.

28.   See "Astronomy Picture of the Day," NASA, March 2, 2009, http://apod.nasa.gov/apod/ap090302.html.

29.   Thanks to Alan Hale for many helpful suggestions on this paragraph, and for informing me of the 1972 Great Daylight Fireball.

30.   According to a recent study, the impactor was a comet, not an asteroid. See Tanya Lewis, "Comet, Not Asteroid, Killed Dinosaurs, Study Suggests," *Live Science*, March 22, 2013, http://www.livescience.com/28127-dinosaur-extinction-caused-by-comet.html.

31.   See Christopher Wills, "Evolution Theory and the Future of Humanity," in *Global Catastrophic Risks*, ed. Nick Bostrom and Milan Ćirković (Oxford University Press, 2008), p. 52.

32.   See Arnon Dar, "Influence of Supernovae, Gamma-ray Bursts, Solar Flares, and Cosmic Rays on the Terrestrial Environment," in *Global Catastrophic Risks*, ed. Nick Bostrom and Milan Ćirković (Oxford University Press, 2008), p. 258.

33.   See Michael Rampino, "Super-volcanism and Other Geophysical Processes of Catastrophic Import," in *Global Catastrophic Risks*, ed. Nick Bostrom and Milan Ćirković (Oxford University Press, 2008), p. 213–214.

34.   See Neil deGrasse Tyson, "We Can Survive Kill Asteroids—But it Won't Be Easy," *Wired*, April 2, 2012, http://www.wired.com/2012/04/opinion-tyson-killer-asteroids/. Interestingly, Sir Martin Rees writes that we're actually "at greater risk from a massive asteroid than from plane crashes." See his *Our Final Hour: A Scientist's Warning* (Basic Books, 2003), p. 89.

35.   See Michael Rampino, "Super-volcanism and Other Geophysical Processes of Catastrophic Import," in *Global Catastrophic Risks*, ed. Nick Bostrom and Milan Ćirković (Oxford University Press, 2008), p. 215.

36.   See Tariq Malik, "NASA Maps Dangerous Asteroids That May Threaten Earth," *Space.com*, August 14, 2013, http://www.space.com/22369-nasa-asteroid-threat-map.html.

37.   Indeed, see "NASA Asteroid Defense Program Falls Short: Audit," September 15, 2014, http://phys.org/news/2014-09-nasa-asteroid-defense-falls-short.html.

38.   See Sarah Titterton, "NASA's Advice on Asteroid Hitting Earth: Pray," *Telegraph*, March 20, 2013, http://www.telegraph.co.uk/science/space/9943048/Nasas-advice-on-asteroid-hitting-Earth-pray.html.

39.   Personal communication.

40.   See "'Extra Funds, No Progress': NASA Watchdog Slams Asteroid-tracking Program," *rt.com*, September 16, 2014, http://rt.com/usa/188036-nasa-watchdog-asteroid-program/.

41.   See Neil deGrasse Tyson, "We Can Survive Kill Asteroids—But It Won't be Easy," *Wired*, April 2, 2012, http://www.wired.com/2012/04/opinion-tyson-killer-asteroids/.

10

Monsters

· · · · ·

Unknown Unknowns

Many riskologists—including myself, in this book—casually identify nuclear weapons as having introduced the first anthropogenic existential risk in human history. Nick Bostrom, for example, writes that "the first manmade existential risk was the inaugural detonation of an atomic bomb."[1] But this is probably incorrect. The Holocene extinction event, for example, likely began in the Pleistocene, when our ancestors started to "overkill" the megafauna. Global warming also began prior to the Atomic Age, and in fact scientists in the 1930s uncovered a warming trend in temperature records dating back to 1865.[2] So there were at least two anthropogenic risks that began to unfold before the Trinity explosion at the Jornada del Muerto (meaning "journey of the dead man") in 1945.

To introduce the topic of this chapter, let's take a closer look at one of the biggest sources of greenhouse gases: *modern transportation*. The story of modern transportation—specifically, of the automobile—is marked by a fair bit of unfortunate irony. At the end of the 1800s, many cities were facing an urban pollution nightmare: they were being overrun with horse manure, urine, and carcasses in the streets. The result was an unbearable odor, major sanitation problems, congestion, gridlock, and swarms of flies (which studies have shown were responsible for "deadly infectious maladies like typhoid and infant diarrheal diseases" in the nineteenth century).[3] And the situation was only getting worse. According to the *Times* of London, nine feet of excrement was projected to cover the streets of London by 1950; on the other side of the Atlantic, an observer suggested that the accumulation of crap would reach the third story of Manhattan's buildings by 1930. The plight of big cities was so dire that the first ever urban planning conference, which hosted delegates from around the world, convened to solve it, but "stumped by the crisis, [they] declared [their] work fruitless and broke up in three days instead of the scheduled ten."[4]

Enter the automobile, which offered a surprising solution to this rapidly worsening public health snafu. "As difficult as it may be to believe for the modern observer," writes Eric Morris, a professor of City and Regional Planning at Clemson University, "at the time the private automobile was widely hailed as an environmental savior."[5] Indeed: no more manure, and the automobile proved to be just as effective a means of transporting goods and people from one place to another. It was consequently adopted *en masse*. What no one foresaw at the time, of course, was that the automobile's internal combustion engine, which converts fossilized plant matter into usable energy by igniting it on fire (thereby releasing $CO2$), would become a major contributor to one of the most significant big-picture hazards of the following century.

Thus, we could say that global warming constitutes an *unintended consequence* of the automobile (although we are fully aware of the connection between automobiles and global warming today). The theorist Langdon Winner defines an unintended consequence as an effect that's "not *not* intended," meaning "that there is seldom anything in the original plan that aimed at preventing them."[6] While unintended consequences have been ubiquitous throughout the great experiment that we call human civilization—a driving force of innovation, in fact—global warming is unique in that it was the first unintended consequence with genuinely *existential* implications.

But it almost certainly won't be the last. If history has taught us anything about purposive human behavior, it's that intended causes proliferate unintended effects. This leads to an absolutely crucial point: *as advanced technologies become more and more powerful, we should expect the unintended consequences they spawn to become increasingly devastating in proportion.* In other words, the future will almost certainly be populated by a growing number of big-picture hazards that were not *not* intended by the "original plan," as it were, and which are significant enough to threaten humanity with large-scale disasters, or even extinction.

Unintended consequences are a kind of "unknown unknown," or a fact about which we are not only ignorant, but ignorant *of* our ignorance. The term "unknown unknown" is perhaps most famously associated with former US secretary of defense Donald Rumsfeld, who, in a 2002 news briefing, explained the concept as follows: "Reports that say that something hasn't happened are always interesting to me, because as we know, there are known knowns; there are things we know we know. We also know there are known unknowns; that is to say we know there are some things we do not know. But there are also unknown unknowns—the ones we don't know we don't know. And if one looks throughout the history of our country and other free countries, it is the latter category that tend to be the difficult ones." (Somewhat amusingly, two years after Rumsfeld's comments, Slovenian philosopher Slavoj Žižek published an article in which he notes that Rumsfeld forgot the fourth possibility: *unknown knowns*, or "the disavowed beliefs, suppositions and obscene practices we pretend not to know about."[7] If there's one *other* category that tends to pose difficulties for the US, I would argue, it's the unknown knowns. Global warming and biodiversity loss could be classified as unknown knowns with respect to the American public.)

I will refer to unknown unknowns somewhat playfully as *monsters*. They constitute an umbrella category, of which unintended consequences are just one type. In addition, the category of monsters also includes (i) phenomena from nature of which we are currently ignorant and which could potentially bring about a catastrophe. For example, there might be risky phenomena lurking about the universe that could destroy our solar system in a whimper or a bang. These might be extremely rare, so we haven't yet observed one occur, and they might require advanced theories in quantum physics to infer—that is, theories we haven't yet developed. Our solar system could thus be obliterated next Friday by an event that we're not only unaware of, but unaware of our being unaware.

Other monsters are (ii) currently unimagined risks posed by future, not-yet-conceived-of technologies. After all, many of the risks discussed in this book were quite unimaginable to people only a few decades ago—and certainly to those living in the 1800s. As we'll explore further in chapter 14, if the development of dual-use technologies continues into the coming centuries, we should expect there to arise brand new existential risk scenarios that, from our current vantage, we can't even *glimpse*. (To be clear, these aren't mere unintended consequences of future technologies—although this will pose a constant hazard—but threats arising from the moral ambiguity of future artifacts' dual usability.) Such scenarios are hidden beneath the horizon of our collective imagination and as such are unknown unknowns. Perhaps if this book were written in 2100, it would be ten times as long and contain a whole different and completely novel set of existential risks.

Finally, (iii) while most of the risk scenarios in this book have been presented as if each constitutes a discrete possible future, virtually all of them can be combined in various ways to produce complex scenarios. Indeed, in some cases, the realization of one scenario will positively *increase* the probability of others occurring. The cliché of a "domino effect" is apt here, as a recent document by the Global Challenges Foundation notes.[8] When two catastrophe scenarios happen simultaneously, their effects can either be *additive* or *synergistic*. Consider, for example, a scenario A that results in 1 billion deaths, and a scenario B that results in 2 billion. In the additive case, of course, the result of A and B together would be 3 billion casualties.

In the synergistic case, by contrast, A and B occurring together could result in *greater than* 3 billion deaths. Imagine that scenario A represents extreme biodiversity loss, but not to the point of initiating an irreversible collapse of the global ecosystem. Now imagine that B involves a regional nuclear exchange between India and Pakistan. It could be that A and B together are synergistically interactive such that the nuclear exchange not only kills 2 billion people but ultimately pushes the global ecosystem past a "critical threshold." The result could be an irreversible collapse of the global ecosystem that causes a total of 6 billion deaths. Many other interactions could be imagined between nanotechnology and biodiversity loss, biodiversity loss and a pandemic, a pandemic and superintelligence, or even a nanoterrorist attack, biodiversity loss, a global pandemic, and a nuclear exchange all in the same few months.

Unfortunately, the possible causal interactions between different risk scenarios unfolding in parallel is a subject that's woefully understudied by contemporary riskologists. Yet it appears likely that—the conjunction fallacy aside—multiple risks happening at once is more probable than only one occurring in isolation.

Varieties of the Unknowable

There are three important distinctions to be made that cut *across* the category of monsters. Let's illustrate the first with an example: imagine that meteorologists predict a hurricane will hit the Florida coast on August 1. Mr. Earl lives in Florida but pays no attention to the news and so doesn't know about the coming storm. Because of this, he goes out for a drive on August 1 and, consequently, gets caught in 130-mile-per-hour winds, torrential rains, and flash floods. In this situation, the hurricane was an unknown unknown for Earl, although it was *knowable* in both practice and principle: Earl *could* have turned the unknown into a known but didn't. In this sense, one might say that the unknown unknown was *person-relative*.

In contrast, the climatic consequences of dumping huge quantities of CO_2 into the atmosphere were unknowable in practice to those who, in the early twentieth century, thought the automobile would ameliorate the problem of urban pollution. The fact is that climatology simply hadn't developed the theories necessary to have accurately predicted that large numbers of automobiles turning fossils into greenhouse gases would threaten later generations with a global catastrophe. As NASA points out in an article on climate change, "Prior to the mid 1960s, geoscientists believed that our climate could only change relatively slowly, on timescales of thousands of years or longer."[9] As the relevant theories were constructed, though, the connection between fossil fuel consumption and global warming became known. In this sense, global warming may have been practically unknowable in 1900, but it was in principle knowable. This type of monster is thus *knowledge-relative*.

The final category has hardly been mentioned in the literature, yet it's extremely important to recognize. It includes monsters that are unknowable in both principle and practice. Recall from chapter 5 that, as Noam Chomsky and other "New Mysterians" have argued, there are intrinsic limitations to the mental machinery behind our eyes: just as the concept-generating mechanisms of the canine brain are inadequate for a dog to ever, in principle, understand the workings of an internal combustion engine, there are phenomena that we'll never, ever be able to grasp, because the concepts needed to understand them forever lie beyond our cognitive reach. How many such concepts—and therefore phenomena—might the human mind be *cognitively closed* to? Who knows. Perhaps an infinite number. It follows that there could be unknown unknowns that are *permanently* unknowable for us. Such monsters are, in this way, *mind-relative*.

Notice that all three distinctions are, terminology aside, ultimately relative to some state of knowledge: the first compares what the individual knows to what the collective as a whole knows, the second compares what the collective knows at one moment in time to what it knows at a later moment, and the third pertains to what our minds are capable of knowing *in principle*.

The second and third categories are most relevant to existential risks. Applying them to a concrete, real-world example: consider the physics experiments being conducted right now in the Large Hadron Collider, the biggest and most powerful particle accelerator in the world. Given our current knowledge of physics—which is both extensive and highly sophisticated, to be sure—there's no reason to think such experiments pose a risk to human existence. But this sense of security may be false. On the one hand, our theories in physics might not be complete, or flawless, enough to anticipate a potential catastrophe. Before the first atomic bomb was detonated, there was "concern that the explosion might start a runaway chain-reaction by 'igniting' the atmosphere."[10] Further investigation revealed that this is physically impossible. Perhaps we're in the exact *opposite* situation with respect to the LHC: perhaps we think there's no risk of an experiment X destroying the earth, but further investigation (say, five years from now) shows that there really is. Perhaps experiment X destroys us in three years, though, before the relevant theory is developed.

On the other hand, there may be side effects of such experiments that we couldn't fathom even if God himself were to descend from the clouds and explain them in perfect detail. The fact is that, as alluded to above, quantum phenomena straddle the Chomskyan puzzle-mystery boundary of human intelligibility. There are many advanced physics concepts that not even the brightest minds can comprehend, such as what the fourth spatial dimension is like. As explained in chapter 5, we grasp such ideas mathematically, not conceptually. Thus, it's entirely possible that a completely inscrutable *unknowable* pops out of the darkness when no one, not even the most brilliant physicists, expects. The result could be something like, to borrow a sentence from Bostrom, "an expanding bubble of total destruction that [sweeps] through the galaxy and beyond at the speed of light, tearing all matter apart as it proceeds."[11]

A dog wandering through the streets of Hiroshima on that tragic day in August 1945 couldn't possibly have anticipated that he was about to be vaporized. Nuclear weaponry is an unknowable unknown in principle for the canine. Perhaps *we* are this dog on the Japanese archipelago, and the bomb heading toward us is a monster wrapped in a mystery. The lesson of cognitive closure is, therefore, that we may be *systematically underestimating the likelihood of disaster*. This is the same lesson as the

Doomsday Argument (which we won't explore here), except that the considerations above are considerably less controversial than the reasoning behind the Doomsday Argument.[12] In other words, the fact that our minds are conceptually limited should lead us to *increase* our prior probability estimates of annihilation, whatever they happen to be. Cognitive closure should make us more pessimistic about the future, however gloomy we might already be.[13]

God Created the Devil

Another unforeseen possibility involves the supernatural. I should point out, right away, that naturalism—i.e., the metaphysical view that there's nothing beyond the natural—is not a dogma of science. It's not an article of faith upon which the scientific enterprise is based, as some religious apologists have charged. Rather, naturalism is best seen as a *theory* about the ultimate nature of reality that's subject to empirical disconfirmation no less than relativity, evolution, and the function of the heart. As such, evidence could arise in the future that calls it into question. If a supernatural deity were to descend from the clouds tomorrow, for example—and if mass hallucination, an extremely well-executed trick by a magician, and so on could be effectively eliminated as explanations—then we'd have very strong evidence for thinking that naturalism is false. The same goes for discovering physical effects without any physical causes in the brain, which would suggest the existence of an immaterial mind interacting with the material brain, as *interactive dualism* (accepted by most religions) posits.

The point is that the best available evidence today strongly suggests that God doesn't exist. Yet it remains entirely *possible* that he does (although see appendix 4 for discussion). There may, in fact, be reasons for thinking a higher power of *some sort* might be lingering in the shadows of reality—reasons unrelated to the Simulation Hypothesis of chapter 6. Consider the Argument from Evil, which many atheists (myself included) see as one of the most powerful arguments against God's existence. It goes as follows: if God is omnipotent (he can at least do anything logically possible) and omnibenevolent (he's morally perfect in every way), then we'd expect the world not to contain every, much less *any*, instance of evil. Yet the world is overflowing with suffering, pain, misery, anguish, despair, hardship, misfortune, and sadness. And, importantly, this evil isn't just the result of human action: there's also "natural" evil associated with "acts of God," such as hurricanes, earthquakes, tsunamis, brain tumors, natural epidemics, and so on. Making matters worse, much—or all?—of this natural evil appears to be entirely *gratuitous*: there isn't any apparent purpose, rhyme, or reason to children dying from leukemia or Tay-Sachs disease. An omnibenevolent God with the power to do anything doesn't appear to be compatible with the world in which we happen to find ourselves.

While this argument usually concludes with the proposition, "And therefore God doesn't exist," this is *not* what it actually supports. All the Argument from Evil shows is that a *perfectly moral* God probably isn't real, given certain facts about our world. The argument leaves open the possibility that a God with a different moral character might exist, such as one who's *morally corrupt*. Indeed, if God were "omnievil," the Problem of Evil would vanish into thin air—it would be solved by being, as it were, dissolved. For various psychologically interesting reasons, no religious tradition has seriously considered the possibility that God, the ultimate creator of the universe, is evil rather than good.

But the proposition that God is omnievil itself encounters a fatal problem, in the form of what we might call the Argument from Goodness. This asks, if God is omnievil, then why is there so much (or, indeed, any at all) goodness in the world? Whereas one would expect *less* suffering if God were morally perfect, one would expect *more* suffering if he were perfectly bad. The evil God hypothesis thus fairs no better than the good God hypothesis.

So where does this leave us? As with so many intellectual debates, it appears that the most plausible hypothesis carves out a middle ground between these extremes. On this view, God is neither good nor evil, but *indifferent*. If this were the case, we should expect to see exactly the sort of world we find around us: one full of inexplicable, pointless, and widespread suffering, as well as meaning, happiness, and prosperity. An indifferent God wouldn't be bothered by earthquakes that deprive children of their parents any more than rainbows and butterflies in a summer meadow. It simply wouldn't, to such a Being, matter either way whether humanity flourishes or perishes, achieves a utopian posthuman state, or goes extinct after an asteroid hits Earth.

I think it's unfortunate that more people—including atheists—haven't taken the existence of a morally indifferent God more seriously. As Richard Dawkins writes in the context of naturalism, accepting that the universe is indifferent "is one of the hardest lessons for humans to learn. We cannot admit that things might be neither good nor evil, neither cruel nor kind, but simply callous—indifferent to all suffering."[14] This idea applies just as well to the context of supernaturalism: it's hard to believe that God might simply have no opinion about us, about the human condition. But an apathetic Artificer could certainly exist, and given the moral properties of our world, a being of this sort appears to be far more probable than an all-good or all-evil deity.

The point is that if God is indifferent, we could be vulnerable to all sorts of unforeseen existential risks. Like someone on a level above us suddenly shutting down our simulation without notice (see chapter 6), an indifferent God could terminate the universe in an instant, on a whim, and without any prior warning. Perhaps—as the philosopher Søren Kierkegaard has speculated—the universe was originally created out of boredom. Perhaps God created it in an attempt to shake his ennui, and this initially worked, but at some point in the future he becomes bored with the tedious spectacle of human affairs. As a result, God destroys the universe and everything it contains. Or perhaps an extremely rare event occurs in the Milky Way galaxy that results in the destruction of the solar system. An all-good God would have, at least presumably, intervened to avert this catastrophe, but an indifferent God lets it happen. One can devise, at will, similar scenarios in which a lack of interest in humanity by the creator of the cosmos leads to completely unforeseen, and maybe even *unforeseeable*, disasters.

In sum, a God-related monster may be improbable from our present vantage, but it appears to constitute a genuinely possible catastrophe scenario: a Being who gives not a damn could either cause, or let happen, an event that wipes humanity out without any warning or moral justification. Continuous with this supernatural possibility is the important thesis that, as the previous section argues, the fact of cognitive closure should lead us to inflate our overall probability estimates of doom, whatever they happen to be. There may be far more existential risks than we (could possibly) realize haunting the cosmos.

Notes

1.   See Nick Bostrom, "Existential Risks: Analyzing Human Extinction Scenarios and Related Hazards," *Journal of Evolution and Technology* 9, no. 1 (2002).

2.   See "The Modern Temperature Trend," American Institute of Physics, February 2015, http://www.aip.org/history/climate/20ctrend.htm.

3.   See Eric Morris, "From Horse Power to Horsepower," *Access* 30 (2007).

4.   The phenomenon of city pollution from horses is well known. This paragraph relied heavily on Eric Morris, "From Horse Power to Horsepower," *Access* 30 (2007).

5.   See Eric Morris, "From Horse Power to Horsepower," *Access* 30 (2007).

6.   See Langdon Winner, *Autonomous Technology: Technics-out-of-Control as a Theme in Political Thought* (MIT Press, 1978), p. 97.

7.   See Slavoj Žižek "What Rumsfeld Doesn't Know That He Knows about Abu Ghraib," *In These Times*, May 21, 2004, http://www.lacan.com/zizekrumsfeld.htm.

8.   See "12 Risks That Threaten Human Civilisation," Global Challenges Foundation, February 2015, p. 20.

9.   See "Taking a Global Perspective on Earth's Climate," NASA, accessed June 17, 2015, http://climate.nasa.gov/nasa_role/.

10.   See Nick Bostrom, "Existential Risks: Analyzing Human Extinction Scenarios and Related Hazards," *Journal of Evolution and Technology* 9, no. 1 (2002).

11.   See Nick Bostrom, "Existential Risks: Analyzing Human Extinction Scenarios and Related Hazards," *Journal of Evolution and Technology* 9, no. 1 (2002). In fact, an explicit aim of such experiments is to discover something *new and surprising*—something that nobody foresaw.

12.   Essentially, the Doomsday Argument goes like this: imagine that you have two boxes in front of you with numbered ping-pong balls in them. Box A contains a total of 100 balls, and Box B only 10. Your task is to guess which box contains 100, and which contains 10. So, you reach into Box A and grab ahold of a ball that, it turns out, is numbered 7. Based on this information, what should you conclude about the boxes? The answer is that you should conclude that Box A has only 10 balls, rather than 100. Now, imagine that you have two hypotheses about the total number of human beings that will ever exist in the cosmos. Hypothesis A posits a total of 200 billion humans before extinction, while Hypothesis B posits 100 trillion. As it happens, the number of humans who've so far been born is about 107 billion in total, according to the Population Reference Bureau (2011). Thus, if your task is to choose between Hypothesis A and Hypothesis B, which do you go for? The answer is, of course, that Hypothesis A is more probable; it's highly unlikely that we find ourselves among the very first humans who've ever been born in the universe. And from this it follows that doom is closer than we'd otherwise expect. A more sophisticated version of the argument merely concludes that however

pessimistic one happens to be, one should be slightly *more* pessimistic: an optimist should be slightly less optimistic, and a pessimist slightly more pessimistic. First proposed by the astrophysicist Brandon Carter in 1983, the Doomsday Argument was developed by John Leslie in his fascinating book *The End of the World: the science and ethics of human extinction.* For more, see George Dvorsky, "Can the Doomsday Argument Predict Our Odds of Survival?" *io9*, April 10, 2013, http://io9.com/can-the-doomsday-argument-predict-our-odds-of-survival-472097460.

13.   These sections drew heavily, and in some cases *ad verbum*, from Phil Torres, "We May be Systematically Underestimating the Probability of Annihilation," IEET, May 27, 2015, http://ieet.org/index.php/IEET/more/torres20150526.

14.   See Richard Dawkins, *River Out of Eden: A Darwinian View of Life* (Weidenfeld & Nicolson, 1995).

11

The Really Big Picture

· · · · ·

The Sun's Future Is Bright

If, by some combination of concerted effort and pure dumb luck, our species manages to survive into the distant future, there are a number of natural events associated with the aging of our Earth, sun, and universe that will pose a succession of existential risks. Indeed, if one were to define existential risks as "causing the extinction of our species," as the Centre for the Study of Existential Risks does, then biological evolution *itself* would constitute an existential risk.[1] This brings us to the really big picture of life. The following offers a snapshot of what we can expect from the universe, zooming from the local to the cosmic.

In the near future, it "appears unstoppable" that a portion of the continental ice sheet making up Antarctica, namely the Amundsen Sea region of the Western Antarctic Ice Sheet, will be lost. This could happen sooner than previously thought, and in fact recent "findings add to a growing body of evidence suggesting that the effects of climate change are outpacing scientific predictions."[2] If this region were to melt, it would be significant, since it "contains enough ice to raise global sea levels by 4 feet." By comparison, the Intergovernmental Panel on Climate Change (IPCC) projects a sea level rise of 1 to 3 feet by 2100, but this estimate doesn't take into consideration the possibility of the Amundsen Sea region crashing into the ocean. More generally, if the entire West Antarctic Ice Sheet were to melt, it "would raise [the] global sea level by about 16 feet," according to NASA.[3] To put this in perspective, a sea level rise of 12 feet would result in the permanent flooding of 98% of New Orleans, 97% of Atlantic City, 73% of Miami, 22% of New York City, and 24% of Boston. This is an effect of short-term climate change that we can expect within the foreseeable future.[4]

The longer-term climatic changes will likely overcome the more immediate effects of industrial activity. We find ourselves in the midst of the quaternary glaciation period, in a momentary warm spell called the *Holocene interglacial period*. Due to changes in properties of Earth like its "axial tilt" (i.e., the degree to which Earth's axis is tilted relative to its orbital plane) and its "orbital eccentricity" (i.e., the degree to which Earth's orbit is an oval rather than a circle), we should expect the climate to begin cooling "any millennium now," with this temperature drop reaching its pinnacle in about 80,000 years. (Note that within the last 1 million years, cold spells have been the *norm*. Our current warm period is very much the exception.) But the anthropogenic, nonnatural phenomenon of climate change could push the next ice age back by thousands of years.[5]

Another cause of environmental change, in the long term, will result from the reorganization of the continents. We know that the crust and upper mantle of Earth is sectioned into various *tectonic plates*, which can move around and collide with each other. About 300 million years ago, all the continents were contiguous—rather than separated by water—forming a supercontinent called Pangaea. This gradually fragmented into two large continents called Laurasia and Gondwanaland. Laurasia became North America, Europe, and Asia, while Gondwanaland became Antarctica, South America, Africa, and Australia. The creation and then separation of supercontinents is a cyclical phenomenon, which means that we can expect another supercontinent to form in the future. Right now, North America is drifting away from Europe at a rate of about 1 inch per year (perhaps a little slower than your fingernails grow). According to one study by the geologist Christopher Scotese, the Mediterranean Sea will vanish in 50 million years as Africa collides with Europe, with a new mountain range in its place as "more and more of the plate [gets] crumpled and . . . pushed higher and higher up, like the Himalayas." Furthermore, the Atlantic Ocean will have gotten larger, and Australia will have shifted north. In about 250 million years from now, a supercontinent will have formed called Pangaea Proxima (formerly, Pangaea Ultima).[6] Simply by chance, this is roughly the same increment of time it takes the solar system to complete a full orbit around the center of the galaxy.

As a result of such geographical alterations, the environments in which our distant descendants (if still on Earth, under a sky full of unrecognizable constellations) live may be radically different from that in which we currently live. New mountain ranges, ocean borders, sea levels, and so on will open up a vast number of new niches for organisms to occupy. Since changes in the environment offer new evolutionary opportunities, we can expect the flora and fauna of the future to be quite different than they are today. If we don't take over our own evolution—as we are now doing, turning ourselves into cyborgs—we can expect natural selection to significantly modify our bodies and brains, resulting in creatures that may be as different from us as we are from the *cynodonts*, a group of animals that lived roughly 260 million years ago and spawned the modern mammal.

Furthermore, the rotational speed of the earth is slowing down, meaning that the days were *shorter* for the dinosaurs and will be *longer* for our progeny. According to one calculation, the day was curtailed by more than an hour some 350 million years ago, and might have been 21.9 hours long 620 million years ago.[7] (Interestingly, the earth's rotation is actually speeding up *right now*, as a result of certain geological phenomena. But this is a temporary trend that will be reversed in the long run.) While Earth spins more slowly (as has already happened to the moon), its axial tilt will undergo changes as well. A paper in *Astronomy and Astrophysics* calculates that in 1.5 to 4.5 billion years, the axial tilt of the earth could reach as high as 89.5 degrees (relative to its orbital plane). The authors conclude that "a very large [axial tilt] in the future of the Earth is a highly probable event."[8] This is significant because it would severely disrupt the seasons, which current life depends upon for survival. Life as we know it could be threatened.

About 2 billion years from now, the Milky Way galaxy (of which our sun is one of 300 billion stars) will begin to collide with the nearest galaxy to us: the Andromeda galaxy. This system, containing about 1 trillion stars, was slightly more than 4 million light-years away when the earth formed about 4.5 billion years ago. Today it's only about 2.5 million light-years away—and quickly approaching. Although this collision won't result in a catastrophic display of celestial fireworks, given how far apart planets and stars are, these galaxies will "eventually settle down into a gigantic swarm of stars buzzing around a common center of gravity," resulting in what T. J. Cox and Abraham Loeb, of the Harvard-Smithsonian Center for Astrophysics, call the Milkomeda galaxy. Both galaxies have supermassive black holes at their centers, and after their merger these black holes may "flare up, becoming active galactic nuclei, feasting on the torrent of new material that was unlucky to enter their feeding zones." It appears as though Earth might be safe from this threat, though, since it may be pushed some

100,000 light-years from Milkomeda's center, into an area safe from the black holes.[9]

While Earth slows down and the Milky Way collides with Andromeda, the sun will enter into the later parts of its life cycle. Like every other star in the universe, it will exhaust its reservoir of nuclear fuel, eventually turning into a cold, dark remnant. Before this happens, though, it will become increasingly luminous. So luminous, in fact, that Earth's surface will rise into the hundreds—even up to ~3,000 degrees Fahrenheit.[10] As the paleontologist Peter Ward and astronomer Donald Brownlee note in their book *The Life and Death of Planet Earth*, "As global temperatures approach, and then exceed . . . 104 degrees F, the planet will begin to die at the equator, and multicellular life will have to migrate poleward." Later, as the average temperature approaches 122 degrees, "wholesale extinction will be taking place on land." At 140 degrees, "the land world will begin to resemble what it once was a billion years ago, before the first animals and plants had evolved." This extinction trend will peak when global temperatures reach 160 degrees: "Except for bacteria, life will have been extinguished." While this occurs, the intense solar heat will also cause the oceans to evaporate into space, leaving behind a desolate, sizzling desert.[11] This will probably happen within the next few billion years.[12]

As the sun continues to age, it will gradually encroach into the solar system, becoming a hugely bloated body intruding on the space of its satellites. Eventually, it will grow so large that it will cover more than half the open sky above us. After consuming the first two rocky planets, Mercury and Venus, our planet will be next on the menu. According to recent calculations by the astronomers Klaus-Peter Schroeder and Robert Smith, this could happen some 7.59 billion years from now.[13] While there's a chance that our planet could escape the sun's gravitational pull, since its mass (and therefore gravitational tug) will decrease as it expands, tidal forces between the sun and Earth "will try to drag the planet inward and downward," thereby foiling its escape plan.[14] In the end, the sun will morph into an engorged red giant, about 256 times as wide as it currently is and 2,730 times as bright.[15] After this, it will become a stellar remnant called a white dwarf, and later a black dwarf that will simply disappear into the dark canopy that forms the backdrop of the universe.

So, big changes await us in the future, as a result of melting ice sheets, natural cycles in Earth's rotation and orbit, continental drift, the slowing down of Earth's rotational speed, changes in Earth's axial tilt, the sun becoming more luminous, and the sun eventually swallowing up the earth. As we'll discuss more in chapter 14, the ultimate strategy for survival in such a universe—one ruled by a dictatorship of entropy—will be to colonize the cosmos.

In with a Bang, Out with a Whimper

But colonizing the cosmos might not be enough, since the cosmos itself will inexorably sink into a lifeless state of frozen chaos: the *big freeze*, resulting from the *heat death*. The story goes like this: about 13.7 billion years ago, the universe began to expand. Counterintuitively, this expansion wasn't *into* space, like the shrapnel of a bomb expanding into air: the cosmos isn't swelling within a larger cavity containing it like a box. Rather, *it's space itself that's expanding*. It follows directly that the Big Bang literally happened everywhere: outside your window, across town, in the next state, where the sun is right now, and at the far edge of the Andromeda galaxy. If cosmic time were to be completely rewound, one would find everything in the universe located at a single point, and nothing beyond it (not even space). The universe is continuing to expand today, and in fact the rate of expansion appears to be increasing (for reasons not clearly understood).

From the Big Bang to the present, the universe has been steadily moving from a state of extremely low entropy, in its initial state, to one of high entropy—where entropy is a measure of how dispersed or concentrated energy is in a system. This is unfolding in accordance with the most "supreme" law of nature, to quote Sir Arthur Eddington, namely the Second Law of Thermodynamics. The Second Law states that entropy will tend to increase over time in isolated systems (i.e., systems unable to exchange either matter or energy with their surroundings). It's this law that accounts for why your coffee gets cold after 20 minutes, you get warm when you exercise, and it's impossible to create a perpetual motion machine (people have tried!).

The march toward a state of *thermodynamic equilibrium*, at which point matter and energy are distributed evenly throughout a system, is a global and unstoppable phenomenon. At present, there are many nonisolated (either "closed" or "open") systems in the universe that are able to temporarily "overcome" entropy by taking in matter and/or energy from the environment to *increase* their own internal order. The earth, for example, is the product of matter and energy being *unevenly* distributed, in the form of oceans, plant life, and human beings. One way it maintains this order (in the thermodynamic sense) is by constantly receiving energy from the sun. Similarly, the way organisms like *Homo sapiens* maintain an extraordinary degree of order within their "internal milieu," a term coined by the French physiologist Claude Bernard, is by taking liquid (water) and matter (food) from the environment, putting it inside their bodies, converting it into energy, and then excreting the remainder. The result is that our internal entropy decreases, but only at the cost of greater entropy in our surroundings.

In the end, though, such rebellions against entropy's reign are destined to lose. The universe will inevitably degenerate into a barren wasteland with a temperature at about 10^{-29} degrees Fahrenheit, according to the theoretical physicists Lawrence Krauss and Glenn Starkman.[16] It will, therefore, morph into a vast and still pond of maximal entropy, one in which life of any kind will be impossible. Indeed, not only will life be impossible, but since information is a disturbance of entropy (i.e., it consists of unevenness, states of order), the entropy death of the cosmos will result in the permanent erasure of every trace, mark, or record of our civilization having ever existed. In the grandest sense, a kind of *cosmic nihilism* frames the human condition: when the eschatology of physics is fully realized, we will have vanished into an oblivion of eternal forgottenness.[17]

Or will we? Some cosmologists have conjectured that there may be an escape from this dismal predicament. One strategy would involve using black holes as portals ("wormholes") to access other regions of the multiverse—that is, neighboring parallel universes that could become a new home. Such universes might be completely different from ours, full of exotic particles obeying unfamiliar laws of nature, or different physical constants. Another possibility would be to create a baby universe into which we could migrate. Yet another involves making a copy of our universe, containing all the things we love and value, and transferring it as a "seed" into another universe. These options are, to be sure, highly speculative; they might best be described as *possibly possible*. Nonetheless, a number of scientists have explored them in print, including the physicist Michio Kaku in his intriguing book *Parallel Worlds* (see the chapter called "Escaping the Universe").[18]

In sum, the universe itself poses an existential risk of the most formidable kind. Just as "thermodynamic equilibrium is the very definition of death" for organisms, to quote the biochemist Bruce Weber in a philosophical article on the nature of life, so too is it the final resting state of the universe.[19] The entropic deterioration of the universe poses the ultimate Great Filter, but it's one we may never have to encounter, given the many risks that dot the road ahead.

Notes

1.   Indeed, if this were the relevant class of organisms, then transhumanism itself would pose an existential risk, since transhumanism is all about replacing humanity with a new and "better" species of *posthuman*. (Yet one wouldn't want to say that transhumanists are working to realize an existential catastrophe.) In the Centre for the Study of Existential Risks' words, "An existential risk is one that threatens the existence of our entire species."

2.   See Chris Mooney and Joby Warrick, "Research Casts Alarming Light on Decline of West Antarctic Glaciers," *Washington Post*, December 4, 2014, http://www.washingtonpost.com/national/health-science/research-casts-alarming-light-on-decline-of-west-antarctic-ice-sheets/2014/12/04/19efd3e4-7bbe-11e4-84d4-7c896b90abdc_story.html.

3.   See "The 'Unstable' West Antarctic Ice Sheet: A Primer," NASA, May 12, 2014, http://www.nasa.gov/jpl/news/antarctic-ice-sheet-20140512/.

4.   See "What Could Disappear," *New York Times*, November 24, 2012, http://www.nytimes.com/interactive/2012/11/24/opinion/sunday/what-could-disappear.html.

5.   See Andrew Revkin, "When Will the Next Ice Age Begin?" *New York Times*, November 11, 2003, http://www.nytimes.com/2003/11/11/science/when-will-the-next-ice-age-begin.html.

6. See "Continents in Collision: Pangea Ultima," NASA, October 6, 2000, http://science.nasa.gov/science-news/science-at-nasa/2000/ast06oct_1/.

7.   See Adam Hadhazy, "Fact of Fiction: The Days (and Nights) Are Getting Longer," *Scientific American*, June 14, 2010, http://www.scientificamerican.com/article/earth-rotation-summer-solstice/.

8.   See O. Neron de Surgy and J. Laskar, "On the Long Term Evolution of the Spin of the Earth," *Astronomy and Astrophysics* 318 (1997).

9.   See Fraser Cain, "When Our Galaxy Smashes Into Andromeda, What Happens to the Sun?" *Universe Today*, May 10, 2007, http://www.universetoday.com/1604/when-our-galaxy-smashes-into-andromeda-what-happens-to-the-sun/.

10. As the astrophysicist Michael Richmond writes, "But, even if the Earth escapes direct contact with the solar atmosphere, it will be doomed nonetheless: the solar luminosity will increase by several thousand as it becomes a red giant, heating the Earth to a temperature of about 2000 Kelvin (about 3000 Fahrenheit). The atmosphere will fly off into space, the oceans will boil away, much of the continents will melt." See "The Future of the Universe," accessed June 17, 2015, http://spiff.rit.edu/classes/phys240/lectures/future/future.html.

11. See Peter Ward and Donald Brownlee, *The Life and Death of Planet Earth: How the New Science of Astrobiology Charts the Ultimate Fate of Our World* (Times Books, 2003).

12. Thanks to the astronomer Robert Smith for some very helpful suggestions in this section.

13. See K-P Schroder and Robert Smith, "Distant Future of the Sun and Earth Revisited," *Monthly Notices of the Royal Astronomical Society* 386, no. 1 (2008): pp. 155–163.

14. See Dennis Overbye, "Kissing the Earth Goodbye in about 7.59 Billion Years," *New York Times*, March 11, 2008, http://www.nytimes.com/2008/03/11/science/space/11earth.html.

15. See Dennis Overbye, "Kissing the Earth Goodbye in about 7.59 Billion Years," *New York Times*, March 11, 2008, http://www.nytimes.com/2008/03/11/science/space/11earth.html.

16. See Michio Kaku, *Parallel Worlds: A Journey Through Creation, Higher Dimensions, and the Future of the Cosmos* (Doubleday, 2005), p. 301.

17. Freeman Dyson has explored ways of surviving this state via hibernation, but Lawrence Krauss and Glenn Starkman argue that Dyson was wrong. See Michio Kaku, *Parallel Worlds: A Journey Through Creation, Higher Dimensions, and the Future of the Cosmos* (Doubleday, 2005), pp. 301–302.

18. See Michio Kaku, *Parallel Worlds: A Journey Through Creation, Higher Dimensions, and the Future of the Cosmos* (Doubleday, 2005).

19. See Bruce Weber, "Life," in *Stanford Encyclopedia of Philosophy*, ed. Edward Zalta (Spring 2015 edition), http://plato.stanford.edu/archives/spr2015/entries/life/. Indeed, this is why disease is often defined as "a breakdown of homeostasis," where homeostasis refers to the process of keeping entropy at bay.

12

The Power of Prophecy

· · · · ·

The Other Branch of Eschatology

Now that we've surveyed a number of secular narratives of gloom and doom, let's take a look at the other side of the field. It's important to understand this half of eschatology because, although the religious and secular branches are conceptually distinct, they overlap in an absolutely crucial way. That is to say, they're causally interactive such that belief in one can actually make the other more dangerous. As we'll see in the following two chapters, it's almost certainly the case that the futurological dogmas of religion are inflating the probability of a secular apocalypse.

As chapter 1 briefly explored, people in all corners of human civilization have long anticipated "the end." This is true today no less than in the past: the eschatological narratives told by the world's religions are still accepted by a large swath of humanity. In some cases, these narratives not only are believed in the abstract, but they also actively shape the behaviors of individuals, groups, and governments in the world. The fact is that *beliefs about the future affect actions in the present*. Let's coin the term "applied eschatology" to refer to any instance in which convictions about the apocalypse influence, shape, or "softly" determine an actor's actions. Applied eschatology has been a driving force in history, and its consequences in recent decades will be the topic of the following chapter. For many, though, it's an invisible force. To actually *see* it being manifested in the world by believers, one needs a rudimentary understanding of the characters, events, and chronologies that comprise the futurological components of the world's major religions. Without such knowledge, the secular riskologist's ability to solve the problem of existential risks will be severely limited.

Thus, the purpose of this chapter is to outline, in detail, the end-times narratives of the two biggest religions in the world: Christianity (in particular, dispensationalism) and Islam (both Sunni and Shia). Not everyone who professes allegiance to these eschatological worldviews will know every aspect of these narratives. But it's better to overshoot than undershoot, if only because many of the eschatological leaders—the most influential Christian and Muslim figures—are familiar with, and *genuinely believe*, such narratives in all their gory minutiae. My hope for the reader is that the following sections can serve as a highly useful resource—perhaps referenced many times, until these convoluted chronologies become familiar—for understanding how religious beliefs about the future are impacting the direction of history.

Abraham's Bosom and the Lake of Fire

The Bible is said to contain "the" capital-t Truth about the nature and purpose of reality. Yet this single book, with its imprecise language, cryptic metaphors, ambiguous allegories, and vague references, has produced a vast array of interpretations, some of which are outright contradictions. It seems odd that a text written by God (through the hands of "inspired" believers) and intended to convey, as they say in court, the truth, the whole truth, and nothing but the truth has given rise to such a rich cornucopia of inconsistent theologies.

One theology that's become quite influential in recent times is called *dispensationalism*. This framework—which itself has a number of variants—gets its name from dividing history into distinct epochs, or "dispensations." It was more or less invented by the Irish preacher John Nelson Darby in the 1800s, although there were similar views floating around at the time.[1] What Darby did is cobble together these ideas into a highly systematic and psychologically compelling theory of the end times. According to Darby's system, history can be divided into seven distinct dispensations: the period before the fall of man, from Adam to Noah, from Noah to Abraham, from Abraham to Moses, from Moses to Christ, from Christ to the Rapture, and finally from the Second Coming to the Great White Throne Judgment at the end of the Millennium. In order, these dispensations are known as Innocence, Conscience, Human Government, Promise, Law, Grace, and the Millennium.

A dispensation is defined by how God choses to relate to his creation and by our differing moral responsibilities toward each other and the

Almighty, "in keeping with how much [we know] about God and His ways."[2] Whereas the only requirement for Adam was "to keep the Garden and not eat of the fruit of knowledge of good and evil," ours is to live by faith and trust in Christ.[3] Each dispensation typically begins with a new revelation for man, and all are destined to end in failure. We're currently living in the second to last period in the grand narrative of creation: the dispensation of Grace, also known as the Church Age. This period began 40 days after Jesus' resurrection (and 10 days after his ascension into heaven) at the first Pentecost of Acts 2:1–47, and it will culminate with the Rapture of the Church—an event that should not be confused with the Second Coming of Christ, also called the *Parousia*.

The dispensationalist narrative of the end times is fairly clear about the chronological sequence of events: at an unknown hour and day in the future, Jesus will return to Earth "with the archangel's call, and with the sound of the trumpet of God" (1 Thessalonians 4:16). But he'll only return *momentarily and not all the way*. This is what dispensationalists refer to as the *Rapture*: it involves Jesus swooping down from heaven and hovering in the clouds while two miraculous events unfold in the "twinkling of an eye." First, all the believers who've died during the Church Age will be resurrected from the grave. And second, all the believers alive at the time will be "caught up" with Jesus in the clouds. (The phrase "caught up" translates as "Rapture" in Latin; the word "Rapture" itself doesn't appear in the Bible.)

In both cases, Christ's followers will receive special new "glorified" bodies. These will be exactly like the body Jesus received upon rising from the dead three days after his crucifixion. The Gospels describe Jesus as being physically absent from the tomb, implying that the very same biological tissues once lifeless in the tomb became magically reanimated. In Luke 24:39, Jesus affirms that his resurrected body is made of "flesh and bones," and the Bible describes his disciples as being able to see him. At one point, in John 20, Jesus exhorts Thomas the Apostle—of "doubting Thomas" fame—to "reach here with your finger, and see My hands; and reach here your hand and put it into My side." Jesus was even able to eat "a piece of baked fish" (Luke 24:35). Yet, curiously, Jesus was also apparently able to walk through walls, as John 20:26 speaks of Jesus entering a room while its doors were shut.

The Apostle Paul, describing the future physique of believers, writes that "the body that is sown is perishable, it is raised imperishable; it is sown in dishonor, it is raised in glory; it is sown in weakness, it is raised in power; it is sown a natural body, it is raised a spiritual body" (1 Corinthians 15:42–44). Interestingly, most Christian traditions interpret "spiritual" here as meaning *not* that we'll be pure souls in the afterlife, perhaps with some sort of ghostly or ethereal body, but that we won't age, succumb to illness, need food and water, or die. Our resurrection bodies will thus be better than the ones we currently have, while nonetheless being *fully physical*. As several theologians put it in a roundtable discussion on eschatology, we'll still be able to play football in heaven.[4]

The idea of bodily resurrection is central to many world religions, including Zoroastrianism, Judaism, Christianity, and Islam.[5] In the dispensationalist narrative, the Rapture coincides with one of *several* resurrection events. All of these, except for one at the very end of time, involve dead *believers* being brought back to life. From the individual perspective, what happens is this: when a believer dies, his or her soul detaches from the body and traverses the cosmos to be "at home with the Lord" (2 Corinthians 5:7), in a fully conscious state. The body from which the soul departed then "returns to the ground it came from" (Ecclesiastes 12:7). This situation, where the body and soul are separated, continues until the Rapture. (Or, for nondispensationalist traditions, it continues until the Second Coming.) When the Rapture occurs, the believer's soul leaves heaven and reunites with its "glorious" new resurrection body. This body-soul coupling—the complete *person*, according to Christian anthropology—then flies back into the clouds to be with Jesus and all the other believers who've been caught up. Everyone then *returns* to heaven for at least seven years (see below), after which everyone, with their bodies and souls, comes *back* to the earth one last time for the final dispensation. But we're getting ahead of ourselves here.

The personal eschatology of *non-Christians* is quite different: upon death, the souls of unbelievers go to "a temporary realm of judgment and condemnation," where they remain, also fully conscious, until an event more than a thousand years after the Rapture.[6] When this future event takes place, the unbeliever's body will be resurrected from the grave and reunited with the soul. But instead of joining Christ in paradise, it will be cast into perdition: the eternal Lake of Fire.

Let's pause for a moment, because already things have gotten complicated. What we've established so far is that the Rapture will occur—no one knows when—and that this event involves Christians from the

Church Age, both dead and alive at the time, being taken back to heaven with glorified bodies. (The dead will have already been in heaven, but not with their bodies.) At this point, one might wonder what the *purpose* of the Rapture is—especially if, as just mentioned, the believers taken into heaven are scheduled to come back to Earth some seven years later. What's the point of this back-and-forth of the faithful?

The curious answer concerns the fact that God has unfinished business with the Jews, his "chosen people." According to dispensationalism, there's a fundamental difference between (i) *Israel*, a community that one is typically *born into*, and (ii) *the Church,* a community into which one is *born again*. While some Christians see the Church as having taken the place of, or superseded, Israel after the Jews rejected Jesus (a theological idea called "supersessionism"), dispensationalists maintain that the Jews are *still* God's chosen people—and God isn't done disciplining them. This is where the Rapture enters the picture. By removing the Church, God frees up the stage of earthly affairs for Israel "to finish transgression, to put an end to sin, to atone for wickedness, to bring in everlasting righteousness, to seal up vision and prophecy and to anoint the most holy" (Daniel 9:24).[7]

This leads us to the next major event in dispensationalist eschatology: the terrifying period of the *Tribulation.* The Bible doesn't tell us how long after the Rapture the Tribulation will begin, but we do know from Daniel (9:24–27) that it will last exactly seven years, with two distinct halves (3.5 years each). The Tribulation will be a period "in which humanity's decadence and depravity will reach its fullness"[8] and great suffering will spread across the globe as "God finishes judging Israel for its sin."[9] Prior to its start (or perhaps during), there will appear sundry "signs of the times," or ominous harbingers of the coming distress. "You will hear of wars and rumors of wars," Jesus says, "but the end is still to come. Nation will rise against nation, and kingdom against kingdom. There will be famines and earthquakes in various places. All these are the beginning of birth pains" (Mark 13:7–8). Jesus points to the rise of false prophets and mass apostasy—a phenomenon occurring throughout parts of the developed world right now—as additional indicators that the end is near.

As the Tribulation begins, a charismatic leader will rise up, calling himself the true Messiah and speaking "proud words and blasphemies" (Revelation 13:5). This demagogue will come to rule a new, reborn Roman Empire, or possibly "a confederacy of ten nations."[10] (Some think he'll become head of the United Nations, others the European Union.)

Eventually, he'll gain "control [of] commerce by forcing people to take the mark of the beast,"[11] an identification code that everyone will be required to have on the right hand or forehead, possibly taking the form of an implanted microchip. He'll also be joined by the "False Prophet," who scripture depicts (metaphorically) "as having horns like a lamb, while speaking like a dragon."[12] Together with Satan, these three figures will form the "unholy trinity," mirroring the Holy Trinity of Father, Son, and Holy Ghost.

But who is this mysterious person with such great power over humanity? The Bible identifies him as the *Antichrist*, or the "beast." Some interpret him as being the first horseman of the Apocalypse, followed by the "horsemen" of warfare, famine, and plague (for a total of four), which together constitute the initial Seven Seals of God's judgment on the world. Many associate the Antichrist with the number 666, a number that's acquired a strong, negative emotional valence over the millennia. Interestingly, though, the oldest extant copy of Revelation 13 (called "Papyrus 115") identifies the number not as 666, but as 616.[13] So perhaps Christians have been fretting about the wrong number for all these years!

The Tribulation officially starts when the Antichrist establishes a covenant with Israel. The first 3.5 years will witness God's judgment unfold with increasing intensity. Wars, famines, pestilence, beasts, and natural disasters will become widespread.[14] There will be exactly 144,000 Jews who come to accept Jesus as their Lord and Savior, spreading the good news of salvation across the world. Many of the unbelievers "left behind" after the Rapture will convert to Christianity, and as a result will be persecuted horribly. Meanwhile, the Jewish temple that was destroyed by the Romans in 70 AD will be rebuilt for the third time on the Temple Mount, and "faithful Jews will once again perform sacrifices."[15] (The Temple Mount is also known as Mount Moriah and Mount Zion, and it's the location of the Dome of the Rock, an iconic Islamic shrine that we'll discuss again later.) The Antichrist will eventually be assassinated, but mirroring Jesus' resurrection from the grave, he'll miraculously return to life. Humanity will come to worship him (in part by force, as those who refuse will be executed), and this will result in unprecedented levels of oppression and discrimination for Christ's followers.[16]

This leads to the second half of the Tribulation, called the "Great Tribulation," a period of spiritual warfare and geopolitical upheaval that will be far worse for the Jews, and everyone else alive at the time, than the

166 • THE END

first half. The Antichrist will invade Israel, as more and more of its citizens acknowledge Jesus as their Lord and Savior. In the process, the rebuilt temple will be desecrated and many will be murdered.[17] The Antichrist will ban all temple sacrifices. In the midst of this chaos, God will unleash two more waves of judgment on the world, the first having been the Seven Seals (of which the horsemen were the first four).[18] The last of the Seven Seals is said to *contain* the next set of judgments, namely the Seven Trumpets. Each of these will inflict horrendous grief upon humanity, and together they will last almost to the end of the Great Tribulation. The Seven Trumpets will bring about a mass famine as one-third of all vegetation is burned. One-third of the sea will turn to blood, killing one-third of all sea creatures and destroying one-third of all ships. One-third of all the fresh water will become undrinkable, and one-third of the sun, moon, and stars will go dark. Finally, a hoard of "possibly demonic" locusts will inflict five months of terrible pain upon those still alive, and one-third of humanity will be destroyed.[19]

When the seventh trumpet blows at the end of the Great Tribulation, a final tsunami of righteous wrath will descend upon the earth: "the Seven Bowls of the wrath of God" (Revelation 16:1). Horrible sores will appear on those who worship the Antichrist, and everything in the sea will die. All "the rivers and the springs of water" will turn into blood, and the sun will "scorch people with fire." The Antichrist's kingdom will be "plunged into darkness," causing people to gnaw "their tongues in anguish" and curse "the God of heaven for their pain and sores." The Euphrates will dry up as "unclean spirits like frogs" emerge from the mouths of Satan and the beast. As the seventh bowl judgment approaches, two "witnesses" (possibly Enoch and Elijah from the Old Testament, neither of whom died) will be murdered, their bodies left in the street for everyone to see. But after 3.5 days, God will breathe life back into their bodies, and "great fear" will overcome those who see them. These witnesses will then ascend to heaven in a cloud while their enemies watch (Revelation 16, 11).

When the seventh bowl is poured into the air, "a loud voice [will come] out of the temple, from the throne, saying 'It is done!'" (Revelation 16:17). A great earthquake will occur and giant hailstones "about one hundred pounds each" will fall from the sky (Revelation 16:21). The Antichrist will gather together an army to overthrow the forces of good. This leads to the very end of the Great Tribulation, marked by the infamous battle of Armageddon and the Second Coming of Christ, when Jesus, riding a white horse, returns not partially as before but *fully and forever*, along with all the believers taken to heaven during the Rapture. The Antichrist and False Prophet are then defeated and thrown into the Lake of Fire, and Satan is shackled in a bottomless pit called the "Abyss."

At this point, God finally fulfills his promises to the Jews: the Palestinian Covenant (Deuteronomy 30:1–10) regarding the *promised land* just east of the Mediterranean—a plot of real estate that includes not only present-day Israel but all of Palestine and Jordan and some parts of Egypt, Syria, Saudi Arabia, and Iraq.[20] This is land that *belongs to the Jews*, according to dispensationalism; no one else has a legitimate claim to it. There's also the Davidic Covenant (2 Samuel 7:10–13) promising Israel an eternal ruler descended from King David (i.e., Jesus), and the Abrahamic Covenant (Genesis 12:1–3) that concerns land, the existence of future generations, a ruler, and a spiritual blessing.[21] Other promises include lifting the curse of creation (Romans 8:18–23) and the eradication of disease (Ezekiel 34:16).[22]

This leads to a judgment of humanity to determine who will gain access to "the kingdom prepared for you from the foundation of the world" (Matthew 25:34). All the nations are gathered before the Son of Man, who "will separate people one from another as a shepherd separates the sheep from the goats. And he will place the sheep on his right, but the goats on the left" (Matthew 25:32–33). The sheep will proceed into the millennial kingdom on Earth while the goats will not. The former group includes not only the previously gathered believers who returned with Christ but all the saints who lived in pre–Church Age dispensations, from Adam to Jesus, as well as those who converted to Christianity *during* the Tribulation. If any of these Christians were killed between the Rapture and the Second Coming, they are here given glorified new resurrection bodies. Interestingly, those who *survived* the Tribulation will continue into the millennial kingdom with their "natural" (i.e., nonglorified) bodies. This will be a crucial detail for the next part of the story, since those with natural bodies will still be able to procreate.

With this judgment, the Tribulation comes to an end. Having brought Israel to Christ and destroyed *all the unbelievers*, the next stage of history is populated, at least initially, only by believers. Here begins the final dispensation of history: the *Millennium*.

According to dispensationalism, the Millennium is a literal 1,000-year period. It's a time of immense happiness and peace between all peoples, nations, and even animals. (As Isaiah 11 states, "The wolf shall dwell with the lamb, and the leopard shall lie down with the young goat, and the calf and the lion and the fattened calf together; and a little child shall lead them.") Deserts will be full of water and sickness will fade into a distant memory of prior dispensations. Jesus will rule the world as a "benevolent dictator" on the throne of David from the city of Jerusalem,[23] and God's relationship with Israel will be renewed.[24] The Temple "will be rebuilt to a previously unseen glory," and everyone will worship there.[25] Life will be wonderful for the inhabitants of Earth.

But a final groan of evil will slowly emerge from the shadows of utopia. All dispensations end in failure, as mentioned earlier, and the Millennium is no exception. Even though Satan is bound in the Abyss, unable to engage in his usual machinations, evil will start to appear as the Christians who converted during the Tribulation and survived into the Millennium with their natural bodies begin to have children. As a result, *new mortals* will be introduced into the world.

While some Millennium-born babies will choose to worship Jesus along with their parents, others will rebel. Toward the very end of this 1,000-year period, Satan will be released from the Abyss, and with greater determination than ever, he'll set out to "deceive the nations which are in the four corners of the earth, Gog and Magog, to gather them together for the war; the number of them is like the sand of the seashore" (Revelation 20:7–8). This desperate insurgency will lead to *the* final battle between God and Satan. Unfortunately for God's nemesis, though, this climactic conflict will end in a quick and decisive victory for God, as fire rains down from heaven and devours the forces of evil (Revelation 20:9). The only remaining member of the unholy trinity (Satan) will then be cast into the Lake of Fire with the Antichrist and False Prophet (who've been there since Armageddon). Evil will finally have been conquered: the war for good will be won for good!

At this point, the earth will once again be populated only by "the righteous," just as it was at the beginning of the Millennium. Those who still have their natural bodies will be given glorified bodies without having to taste death (just like those who were alive during the Rapture), and the "last vestiges of the curse on creation" will be lifted once and for all.[26] Meanwhile, those who rejected Christianity during the Church Age and Millennium will be resurrected, their souls having been temporarily stationed in the place of judgment and condemnation mentioned earlier. This is the *one and only resurrection* involving unbelievers.

Although their fate is already sealed, these heathens, reprobates, atheists, and infidels will be judged by Jesus at the Great White Throne Judgment, an event involving only non-Christians. A couple of books will be opened. One of them, the Book of Works, will contain a detailed record of what every sinner has ever thought, said, and done. The damned will receive punishments according to the life records written in this book. But their ultimate fate will be determined by whether their names appear in a second book: the Book of Life. If your name is included, then heaven is yours. If not, then your fate is eternal torture amid fire and brimstone. In this way, just as believers experience a "second birth" when they convert to Christianity, unbelievers will undergo a "second death" upon being resurrected and then cast into hell.[27]

This leads to the terminus of God's grand plan for the world, at which point God will completely destroy and then recreate the universe. As 2 Peter 3:10 puts it, "The heavens will pass away with a roar, and the heavenly bodies will be burned up and dissolved, and the earth and the works that are done on it will be exposed." The result will be "a New Heaven and a New Earth for believers to live in for eternity."[28] This new world *is* paradise—it's a literal *heaven on Earth*. It will contain a New Jerusalem in the expanded territory of Israel and will be completely free of sin, suffering, and death. "He will wipe away every tear from their eyes," Revelation 21:4 states, "and death shall be no more, neither shall there be mourning, nor crying, nor pain anymore, for the former things have passed away." Eternal peace has arrived. The struggle is over.

This concludes the powerful, imaginative, and quite terrifying eschatology of dispensationalism.

The Most Beneficent, the Most Merciful

The eschatology of Islam shares a number of similarities with Christian, Jewish, and Zoroastrian eschatology (see appendix 2 for more). In this section, we'll start with a quick look at the basics of Islam, and then turn to how Sunnis and Shias think the end will unfold.

Islam, meaning "voluntary submission to God," was founded by Muhammad, an illiterate merchant born in 570 AD. At the age of 40, while retreating in a cave outside Mecca (in modern-day Saudi Arabia), Muhammad received his first revelation from the angel Gabriel (or *Jibril*). For the next 22 years, until his death, Muhammad continued to receive revelations, which were recorded and compiled into Islam's holy book, the Qur'an (or Koran), which consists of 114 suras, or chapters. This is considered to be the literal, immutable, infallible Word of God (or *Allah*, which literally translates to "the God"). In addition to the Qur'an, there's a large corpus of texts in which the deeds and sayings of Muhammad, called the *sunnah*, have been recorded. These texts are referred to as the *hadith*. Each hadith can be ranked along a spectrum of authenticity, with "weak" and "strong" (defined according to a set of scholarly criteria) at the extremes. The hadith are important because Muslims believe that Muhammad was not only a prophet of God and his messenger but also a man of exemplary spiritual character.

Muhammad was born in Mecca, now considered to be the holiest city in Islam. As mentioned in chapter 1, the Grand Mosque (or Masjid al-Haram), the largest in the world, is located in Mecca. It's built around the Kaaba, a sizable cuboid structure that Abraham himself is said to have constructed with the help of his son Ishmael. The second holiest city is Medina, where the first Muslim community was established by Muhammad after leaving Mecca. Jerusalem is the third holiest city, and is home to the very recognizable shrine called the Dome of the Rock, with its glimmering golden dome, built on the Temple Mount in Jerusalem's Old City—the exact same location, you may recall, that dispensationalists believe the Third Temple will be erected and later destroyed during the Tribulation. Jerusalem was also the destination of a mystical expedition that Muhammad went on one night, known as the Night Journey. Flying on a magical, white, winged steed, Muhammad ventured from the Kaaba, near to where he was sleeping, to the Temple Mount.[29] He then climbed a ladder through the seven stages of heaven, meeting a number of past prophets along the way, such as Abraham, Moses, John the Baptist, and Jesus (or *Isa*).

Just like Judaism and Christianity, Islam traces its lineage back to a single person: Abraham. The story goes like this: Abraham was married to Sarah, but she was unable to produce a child. Consequently, Sarah offered her Egyptian handmaiden, Hagar, to Abraham as a second wife, and Hagar later gave birth to a son named Ishmael. God then gave Sarah, at this point 90 years old, a child named Isaac. This is an important detail because whereas Jews and Christians trace their lineage back to Abraham through Isaac, Muslims trace it back through Ishmael. The genealogical convergence here, through the sons of Hagar and Sarah, is why Judaism, Christianity, and Islam are considered "Abrahamic religions."

It should be emphasized that Muslims see their religion as *continuous* with Judaism and Christianity, not unlike the way Christians see their religion as the Old Testament *plus* the New Testament. Indeed, Muhammad's Night Journey is significant in part because it affirms this relation of continuity.[30] According to Islam, there have been many revelations handed down by God through past prophets, including Adam, Abraham, Ishmael, Isaac, and Jesus. Muhammad was the last—indeed, the culmination—of this long succession of prophets. The Qur'an thus constitutes the complete and final revelation of God to humanity. It's also intended to *correct* the traditions established by past prophets, which Muslims believe became corrupted and compromised by extraneous doctrines or deliberate interpolations that accumulated over time.[31]

For example, elevating Jesus to the level of God (as in the Holy Trinity: the Father, the Son, and the Holy Ghost) is to commit the unforgivable sin of idolatry, called *shirk*. Muslims believe that Jesus was a *prophet*, but not God. (This is roughly the same view they hold of Muhammad, and it's why drawing the prophet is prohibited: doing so could lead to worshipping Muhammad as if he were a God, the way Christians worship Jesus.[32]) Furthermore, the Christian belief that Jesus died on the cross is seen as a corruption of the true story. In Islam, there's no doctrine of original sin, no Fall of Man after Adam ate the apple in the Garden of Eden, and therefore no need for Jesus to atone for our sins. God thus took Jesus into heaven and, on one interpretation, made it *look* like he'd been crucified. He did this, some have speculated, by turning Judas into a substitute; another view is that Simon of Cyrene, who the Gospels describe as being forced to carry Jesus' cross, was magically transformed to look exactly like Jesus and consequently was executed in Jesus' place.

Perhaps the most significant feature of Islam is its *uncompromising monotheism*, which affirms the essential *oneness* of God. The Christian notion of the Holy Trinity being three completely distinct persons who are also completely identical in essence is, for Muslims, incoherent and absurd. How is that possible? It seems to defy logic itself. (Indeed, the word "shirk" is sometimes translated as "polytheism.") It's this emphasis on the singularity of God's essence that constitutes the most fundamental doctrine of Islam, called *Tawhid*. The proclamation of this belief, or *shahada*, makes up the first of the Five Pillars of Islam, which lie at the heart of Islam (both Sunni and Shia). Whereas being a Protestant Christian depends almost entirely on what one believes, the oneness of God and Muhammad being his messenger are the only beliefs one must have to be a Muslim. According to the scholar John Esposito, the stress in Islam is on *acts* ("orthopraxy") rather than beliefs ("orthodoxy").[33] In addition to shahada, the other Five Pillars are praying five times a day facing the Kaaba, giving to the poor, taking a pilgrimage to Mecca at least once in one's life if possible (the *Hajj*), and abstaining from eating, drinking, and having sex during the daylight hours of Ramadan, the ninth month of the Islamic calendar (and the month during which Muhammad received his first revelation).

Just as Judaism has spawned the Orthodox, Conservative, and Reform sects over time, and Christianity its Protestant, Catholic, and Eastern Orthodox denominations, the religion of Muhammad has produced two distinct branches: Sunni Islam and Shia Islam. The latter is a minority tradition, consisting of between 10% and 15% of the world's 1.6 billion Muslims. The historical split between Sunnis and Shias—which is important for understanding many current affairs, as we'll discuss—resulted from an early disagreement about who should lead the Muslim community after Muhammad's sudden death in 632.[34] The majority of Muslims at the time believed that the leader, or *caliph*, of the Islamic government, or *caliphate*, should be chosen by the community. A smaller group held that not long before his death, Muhammad gave the leadership over to his cousin and son-in-law, Ali (who was married to Muhammad's daughter, Fatimah). As Ron Geaves notes in his book *Islam Today*, according to this view, Muhammad "came to know that Ali was his successor when it was revealed to him on the Night Journey," and he's reported to have said: "He, of whom I am the patron, of Him Ali is also the patron."[35] The community of Ali supporters became known as the "shi'at Ali," meaning "the party of Ali," which is shortened today to "Shia."[36]

A succession of four caliphs followed Muhammad's death: Abu Bakr, Umar, Uthman, and Ali. These are considered the "Rightly Guided Caliphs" by Sunnis, while Shias hold that the first three were illegitimate usurpers of the caliphate. This led to a great deal of violence, and consequently "two of the earliest caliphs were murdered."[37] Eventually, Ali became caliph, accepted by both Sunnis and Shias, but he was later murdered as well. The position of caliph was then transferred not to Ali's son Hasan, as the Shia believe it should have been, but to the governor of Syria, Muawiyah. After Hasan's death in 670 AD (possibly as a result of poisoning), Shia believe that his younger brother, Hussein, should have become caliph, but instead Muawiyah gave it to his own son, Yazid.[38] (Muawiyah died exactly ten years after Hasan.) It was this nepotistic transfer of power that turned the caliphate into the first of several hereditary dynasties: in this case, the Umayyad dynasty. Hussein, though, refused to pledge allegiance to Yazid and consequently set out to challenge him. But on the way, he found himself surrounded by the far more powerful forces of Yazid. The resulting massacre by Yazid's army occurred in the Iraqi city of Karbala: Hussein was brutally assassinated and his remains mutilated, a trauma that still haunts the collective memory of Shia today.[39] This is the infamous Battle of Karbala, and it's commemorated during the first month of the Islamic calendar, Muharram.

According to the Sunni tradition, the caliphs are mere mortals—men with exceptional leadership skills but no supernatural abilities. The Shia view is quite different: they refer to the legitimate Muslim rulers, the caliphs, as *Imams*, where this title designates not only a political leader but a spiritual one as well. The Imams are, therefore, "the true inheritors of Islam, the direct line of the Prophet, replete with . . . special powers inherited through his bloodline that permitted them to know the inner secrets of the Qur'an."[40] As one of the leading scholars of Islam, David Cook, puts it, Imams have "exclusive knowledge of the past and future," "access to interpretations of the Quran to which no one else is privy," and "a unique connection with God" (that is, as a continuation of "the prophetic experience of Muhammad").[41] There are several traditions within Shia Islam, but the dominant one maintains that, from Ali to the present, there have been exactly 12 Imams. Those who subscribe to this view are called *Twelvers*, and they make up a majority of Muslims in Iran and Iraq and have a notable presence in Lebanon, Pakistan, India, Bahrain, Afghanistan, and even eastern Saudi Arabia.

Before moving on to what Sunnis and Shias believe will happen in the end, we need to examine the personal eschatology of Islam—or what happens to us as individuals after we die. Recall that Christians believe the fully conscious souls of believers travel to be with Jesus until the Rapture or the Second Coming, at which point the body is resurrected and reunited with the soul. Bodily resurrection is, as mentioned, also a central dogma of Islam. Roughly speaking, what happens is this: according to the Qur'an and hadith reports, either God or an "Angel of Death" takes the soul from the body. The soul remains contained in the grave for the first night after death. Two angels then approach the soul and ask it "about its God, its prophet, and its scripture."[42] Answering correctly results in a temporary journey "through the heavens into the presence of God," after which the soul is returned to its earthly resting place; answering incorrectly results in no such expedition. In both cases, the soul (then) stays put until the body is raised from the grave at the very end of time. In other words, the time between death and the resurrection is spent in the grave. For the heaven-bound, though, this period "passes quickly and pleasurably," while "for the damned, slowly and painfully."[43]

It's extremely difficult to piece together a single chronological narrative of end-times events according to Islamic sources. The tradition of apocalyptic literature in Islam "is not unified or codified in the manner of the Islamic legal tradition," as Cook notes.[44] The Qur'an and hadith (of which there are some 143 dedicated to eschatological themes) present an incomplete, fragmented picture of the world's ultimate conclusion, and Shia expectations differ fundamentally from those of the Sunnis. As Marcia Hermansen observes, "The Islamic concept of time is frequently less linear than that of the Christian and Jewish traditions,"[45] and Jean-Pierre Filiu notes that the Qur'an itself "provides few clues regarding the apocalyptic calendar."[46] Given such caveats, the narrative below attempts to approximate the more popular accounts of eschatology taken seriously by many believers today. (This is precisely what we did earlier, focusing on dispensationalism not because of its academic respectability but because of its prominence in American religious culture.) Significant traces of the following account can be found in the writings of contemporary Muslim apocalypticists, as well as the scholarly works of people like Cook, Filiu, and the Islamic scholar and cleric Yasir Qadhi.[47]

To begin, Sunnis and Shias both identify a man called the *Mahdi* as a central figure of the eschaton, although his status in each tradition is quite different. The Mahdi is a messianic figure whose purpose is to unite the Muslim community (*ummah*) and usher in the final events leading to the Day of Resurrection, Judgment, Reckoning, or the Last Hour. While the Mahdi isn't mentioned in the Qur'an, there are numerous hadith that discuss his attributes and future actions; some of these hadith are strong, although many more are weak. According to one such account, the Mahdi will be a descendant of Muhammad; his name will be Muhammad (just like the prophet) and his father's name Abdullah (just like the prophet's father).[48] He "will have a broad forehead [and] a prominent nose," which Qadhi suggest could either be because his hairline starts further back on his skull or because he has low eyebrows.[49] Furthermore, it could be that he won't always have been a pious Muslim but will undergo some sort of epiphany "during a single night."[50] This will turn him into a fervent believer. Once the Mahdi appears, he "will rule for seven years," although some traditions specify his leadership as lasting for five or nine years.[51] While many of his physical characteristics will differ from Muhammad's, he'll share the same *spiritual* qualities as the Prophet, such as being morally upright, merciful, and forgiving.[52]

An approximate chronology of the Mahdi's emergence, from the Sunni perspective, goes like this: first, the Mahdi will not *want* to be the Mahdi. Nor will he initially be aware of his status *as* the Mahdi. Fleeing from Medina to Mecca, on one view, he'll seek refuge in the Kaaba because of growing *fitna*, or strife, in the world.[53] Some people in Mecca will recognize him as being the Mahdi and consequently will force him out of his home, where they'll give allegiance (*bay'ah*) to him between the Black Stone (a cornerstone of the Kaaba thought to date back to Adam) and the Station of Abraham (an imprint of Abraham's foot that was left while building the Kaaba with Ishmael).[54] Meanwhile, as one hadith puts it, "an army would be sent to fight (and kill) them." But a miraculous event will prevent this marauding army from reaching the Mahdi and his followers, who will be more or less defenseless at this point, having no weapons with them. As the army approaches Mecca, the earth will suddenly open up and swallow it whole. This opening of the earth will, in fact, swallow up everyone in the vicinity, including innocent travelers who have nothing to do with such martial expeditions. These people, Muhammad is reported to have said, will "be raised on the Day of Resurrection on that basis of [their] intention."[55]

It's this event that will offer unambiguous proof to Muslims that the Mahdi has arrived. Adding to the evidence, another army carrying black flags (or banners) will emerge from the East, possibly from Khorasan, to protect the Mahdi in Mecca. (Khorasan is a historical region that included parts of Afghanistan and Central Asia.) Such events will indicate to the entire Muslim world that they should give allegiance to this man, the Mahdi, and flock to him even if one must, as a weak hadith states, "crawl over snow."[56] The Mahdi will subsequently become the undisputed leader of the Muslim world, filling "the earth with equity and justice as it has been filled with oppression and tyranny." He will be a caliph who distributes wealth so generously that he doesn't even count money.[57] A period of joy and abundance will ensue.

But this is not the end, only the beginning. According to tradition, the apocalypse will be precipitated by a number of "signs." These can be divided into the Lesser and the Greater. The Lesser Signs include trends like the lifting of (religious) knowledge, the spread of ignorance, an increase in wars, the proliferation of evil, and earthquakes. There will be widespread unrest, and murder will become "frequent." As a famous hadith puts it, the Last Hour won't "arrive until people build buildings that are too high and until whoever passes by a tomb says: I wish to Allah that I were in the place of the one buried there."[58] Many of these signs are all around us today, as Muslim apocalypticists are quick to point out, although they're vague enough to apply to just about any age. While the Lesser Signs are relatively mundane—albeit undesirable—in nature, the Greater Signs are extremely unusual, supernatural events. There are, according to one hadith, ten of these signs in total.[59] What's important to note for now is that the Mahdi is *neither* a Lesser nor Greater Sign. Rather, Qadhi asserts, he constitutes the *link* between these two categories; he is the agent of transition from the worsening conditions of the world to the era of overtly supernatural phenomena, which will unfold very quickly once it commences.[60]

The first of the Greater Signs will be the appearance of the *Dajjal* (or Antichrist). The Dajjal will be, according to one hadith, "one eyed, his right eye . . . protruding out [like a] grape."[61] He's also "usually said to be a Jew who will appear from the east at the head of an army."[62] His reign of terror on Earth will last for "forty days, one like a year, one like a month, one like a week, and [the] rest of his days like yours."[63] Qadhi interprets this passage as meaning that, using a little numerological calculation, the Dajjal will be around for a year and two and a half months, presumably toward the end of the Mahdi's seven-year rule.[64]

The Dajjal won't enter the picture, though, until a great battle between the Muslims and the "Romans" (often interpreted as the US by contemporary apocalypticists) occurs in a small town of northern Syria. As a hadith puts it, "The Last Hour would not come until the Romans would land at al-A'maq or in Dabiq. An army consisting of the best (soldiers) of the people of the earth at that time will come from Medina (to counteract them). When they will arrange themselves in ranks, the Romans would say: Do not stand between us and those (Muslims) who took prisoners from amongst us. Let us fight with them." The "prisoners" referred to here are Roman captives (or defectors?) who've converted to Islam. But the Muslims will refuse, saying, "Nay, by Allah, we would never get aside from you and from our brethren that you may fight them." The two armies will then engage in battle, by the end of which one-third of the Muslim army will flee, one-third will perish, and one-third will be victorious. Those who have survived will, as a result of this extraordinary test, never be challenged again, while those who fled will never be forgiven by God, and those who died will be labeled "excellent martyrs."

The victorious army will then proceed to Constantinople (now called Istanbul, in northwestern Turkey) to conquer it as well.[65] This won't be an ordinary victory, though. Seventy thousand Muslims will land there, but "they will neither fight with weapons nor shower arrows" upon the city. Rather, they will only shout: "*There is no god but Allah and Allah is the greatest!*" As a result, "one side of [Constantinople] would fall," possibly the "part by the side of the ocean." A second time, the army will exclaim, "*There is no god but Allah and Allah is the greatest!*" and "the second side would also fall." Finally, "the gates would be opened for them and they would enter."[66] This series of events is rather perplexing given that Constantinople was conquered by Ottoman Muslims in the fifteenth century. "Some solve this difficulty," Cook notes, "by seeking to deny that the original conquest in 1453 was actually accomplished by (true) Muslims so that it can be 'reconquered' at the end of the world."[67]

This is where the Dajjal enters the narrative. According to one hadith, after the army at Constantinople "[hangs] their swords by the olive trees" and begins "distributing the spoils of war (amongst themselves)," someone will cry: "The Dajjal has taken your place among your family." Many will then run home, only to discover that this was a lie perpetrated by Satan. After this happens, the army will reconvene in Syria, and while preparing for battle against the Dajjal (protected from his spells by reciting a sura of the Qur'an), the Muslims will stop for the *Fajr* (morning) prayer.[68] It's here that the second of the Greater Signs will take place: the descent of Jesus "at the white minaret to the east of Damascus" (the capital of Syria), being supported by two angels, one on each side. As Cook points out, "Although such a return is not mentioned specifically in the Qur'an, it is strongly implied by the wording of Qur'an 3:55," which states, "Allah said, 'O Jesus, indeed I will take you and raise you to Myself and purify you from those who disbelieve and make those who follow you [in submission to God alone] superior to those who disbelieve until the Day of Resurrection. *Then to Me is your return*, and I will judge between you concerning that in which you used to differ'" (italics added). According to one hadith, when Jesus lowers his head, "there would fall beads of perspiration from his head, and when he would raise it up, beads like pearls would scatter from it. Every nonbeliever who would smell the odor of his self would die and his breath would reach as far as he would be able to see."

Jesus will then chase down the Dajjal, along with 70,000 Jews who are with him,[69] until they reach the "gate of Ludd"—modern-day Lod, in Israel. (Some accounts have it that Jesus descends to the earth over Jerusalem rather than near Damascus.) At this point, Jesus will defeat the Dajjal, showing his fellow Muslim fighters the Dajjal's blood on his spear. He will also, as several hadith put it, "kill the pigs" and "break the cross," the latter of which is especially significant, since it will prove that Christianity is a corruption of the Truth.[70] The Jews who've not converted to Islam will be exterminated here as well; to quote a particularly disturbing hadith, "The Last Hour will not come until the Muslims fight against the Jews and kill them. And when a Jew will hide behind a [stone] wall or a tree, the wall or tree will cry out: 'O Muslim! O servant of Allah! There is a Jew behind me.' And [the Muslim] will come to kill him."[71] Incidentally, it's not clear what happens to the Mahdi from here out, as he suddenly disappears from the prophetic narratives. Perhaps he dies naturally or is killed, having completed his eschatological mission on Earth.

Jesus will then rule for 40 years, according to some accounts. Having now killed the Dajjal, he "will comfort the surviving and sorely tested Muslims," leading "them into the calm haven of Tur" (a mountain, possibly atop the Mount of Olives on the eastern side of Jerusalem).[72] This is followed by one of the more catastrophic of the Greater Signs of the apocalypse: the accursed, bellicose peoples of Gog and Magog will "swoop down from every mound" and "prevail over the earth."[73] These peoples, as Cook observes, "act as killing machines designed to destroy the entire world . . . for no apparent reason."[74] Not only will they be violent, but they'll consume the earth's resources to the point at which whole bodies of water will go dry after they drink from them. As one hadith states, "Muslims will flee them until the remainder of the Muslims are in their cities and fortresses, taking their flocks with them." This is where something miraculous happens: God utterly decimates Gog and Magog by sending "a worm like the worm that is found in the noses of sheep [to] penetrate their necks." Consequently, they "die like locusts, one on top of another." As a result, the next morning "the Muslims will not hear any sound from them, and they will say: 'Who will sell his soul for the sake of Allah and see what they are doing?' A man will go down, having prepared himself to be killed by them, and he will find them dead, so he will call out to them: 'Be of good cheer, for your enemy is dead!'" As another hadith puts it, the Muslims "will then come down and they will not find in the earth as much space as a single span which would not be filled with their corpses and their stench."

As the final pages of history turn, more Greater Signs appear. For example, the sun will rise from the west rather than the east. This event is significant for humanity because it marks the closing of the door of repentance. To quote a hadith, "Faith will not benefit any soul that did not believe before or earn anything good through its faith." The fate of each person's soul will thus be forever sealed. Another major sign is the so-called Beast of the Earth. As the Qur'an states, "And when the Word is fulfilled against them (the unjust), we shall produce from the earth a beast to (face) them: He will speak to them, for that mankind did not believe with assurance in Our Signs." According to a (weak) hadith, the beast will have the ring of Solomon and the staff of Moses. Its purpose, on some accounts, will be to differentiate the community of believers from that of the nonbelievers (or Kafir). The Beast will then "brighten the face of the believer, and stamp the nose of the disbeliever with the ring, such that when the people gather to eat, it will be said to this one: 'O believer!' And to that one: 'O disbeliever!'"

One of the last Greater Signs is the Pleasant Wind that, after the destruction of Gog and Magog, "will seize [believers] beneath their armpits and will take the soul of every Muslim, leaving the rest of the people fornicating like donkeys." The Kaaba will, apparently, be destroyed by a "thin-legged" leader from Ethiopia, named Dhul-Suwayqatayn, and a great fire will eventually result in "the gathering together of humanity . . . to the place of the Final Judgment."[75] As the end inches nearer, an angel named Israfil "will sound the horn signaling the arrival of the Hour and will read from the guarded tablet that which is written concerning the lives of all who will then be brought to judgment."[76] The trumpet will be sounded either once (Sura 69:13) or twice (Sura 39:68), although some traditions combine these to give a total of three trumpet sounds.[77] To quote the Qur'an at length:

Then when the Horn is blown with one blast, and the earth and the mountains are lifted and leveled with one blow—then on that Day, the Resurrection will occur, and the heaven will split [open], for that Day it is infirm. And the angels are at its edges. And there will bear the Throne of your Lord above them, that Day, eight [of them]. That Day, you will be exhibited [for judgment]; not hidden among you is anything concealed. So as for he who is given his record in his right hand, he will say, "Here, read my record! Indeed, I was certain that I would be meeting my account." So he will be in a pleasant life—in an elevated garden, its [fruit] to be picked hanging near. . . . But as for he who is given his record in his left hand, he will say, "Oh, I wish I had not been given my record and had not known what is my account. I wish my death had been the decisive one." . . . [Then God will say], "Seize him and shackle him. Then into Hellfire drive him. Then into a chain whose length is seventy cubits insert him."

Although the judgment itself has already concluded at this point, the Qur'an suggests that people must cross a bridge over hell. Whereas the faithful will "move easily and swiftly across a broad path, led by the members of the Muslim community and first of all by the Prophet himself," the unbelievers with "neither faith nor good deeds to their credit [will] find that the [bridge] has become sharper than a sword and thinner than a hair, and that their fall from it signifies an inescapable descent into the Fire of everlasting punishment."[78] The Qur'an isn't exactly shy about the nature of hell. It will be a place of eternal torture, one full of unquenchable fires "whose fuel is people" (Sura 66:7). Heaven, on the other hand, will be a place of tall buildings, flowing rivers, abundant food and drink, shade, garments of fine silk and brocade, thrones, and plenty of women with large, beautiful eyes. This is a basic outline of eschatological narratives in the Sunni tradition.

The Twelver Shias have a rather different understanding of the end. According to them, "beliefs in the coming of the messiah focus entirely upon the figure of the Twelfth Imam." This Imam's name is Muhammad ibn al-Hasan (where "ibn" means "son of" in Arabic[79]), and in 874 AD, at the age of five, he disappeared, entering into *the Occultation* for "self-protection." (This is why he's called the "Hidden Imam.") The Twelfth Imam has thus been alive for over 1,000 years, living (possibly) in a cave in Iraq or, as many Iranians believe, inside the shrine at Jamkaran, a village in the central region of Iran.[80]

Twelvers believe that this Imam *is* the Mahdi, and they see his role in the eschaton as being far more central than Jesus'. His reappearance will occur during a time of great strife for the Muslim world. According to some accounts, multiple eclipses during the holy month of Ramadan and "ravaging swarms of locusts" will precede the return of the Twelfth Imam. Fighting will destroy Syria, "a rain of reddish fire will fall on Baghdad and Kufa," and the sun will rise from the west, just as occurred in the Sunni narrative.[81] One of "five signs" outlined by Shia traditionalists is that a tyrannical Arab ruler called the *Sufyani* (the sworn enemy of the Mahdi) will come from Syria and "brutally oppress the Shiite peoples."[82] There will be multiple invasions of Muslim lands by the Byzantines (or eastern Romans), which many Muslims today interpret as the US and its allies. Indeed, Cook observes that "it is widely accepted that [the] US invasion of Iraq in 2003 fulfilled these predictions."[83] As happened in the Sunni narrative as well, an army will be sent to kill the Mahdi, but it will be miraculously swallowed whole by the earth between Mecca and Medina.[84]

Once the Mahdi reemerges, he will "manifest himself as the Master of the Age . . . and his return [will] mark the end of the cycle of creation."[85] But, as Filiu notes, the Mahdi will also be "the Master of the sword . . . for he will mercilessly punish the enemies of Islam."[86] In doing this, the Twelfth Imam will demolish mosques, destroy the religious scholars of the Sunnis (the *ulama*) for their theological heresies, and ultimately "take vengeance upon those Sunni Muslims who have opposed the rights of the family of the Prophet Muhammad to rule."[87] According to Qadhi, the Mahdi may even oversee the resurrection of all the Sunni rulers only to crucify and torture them, perhaps killing them over and over again as punishment for their supposed evil.[88] Such apocalyptic prophecies are consistent with the long-standing feelings of persecution and oppression felt by the Shia minority at the hands of "corrupt" Sunni rulers.

The Twelfth Imam will bring justice to the world. He will, in Cook's words, "establish a messianic state that will encompass the world."[89] Beyond this, his eschatological importance is underlined by the assertion that *he*, rather than Jesus, will eventually kill the Dajjal. On the Shia account, the Mahdi will lead an army, called the "Mahdi Army" (a name to keep in mind for the next chapter), to the Byzantine city of Constantinople, at which point the Dajjal will appear and Jesus will intervene; but the act of defeating the Antichrist will be given to the Twelfth Imam. In the end, he "will set Islam back on the straight and true path from which it had strayed."[90] This leads to the ultimate climax of history: the Final Judgment of humanity by God.

As Filiu notes, the overarching structure of Islamic eschatology combines messianism and millenarianism in a rather unique manner. In Islam, the *messianic* hope of a Muslim future is bound up with the Mahdi, a figure whose function in the end times is to restore Islam as the force it once was. This restoration is of *millenarian* significance in that it results in a series of catastrophic battles that ultimately transform the world forever.[91] We can contrast this with Christianity, which is unique in that it basically takes the narrative of Jewish eschatology—the expectation of a Messiah who will save humanity from evil—and places it inside a *larger* narrative in which the Messiah, Jesus, makes an initial appearance, ascends to heaven, and then returns millennia later to implement a Millennial Kingdom on Earth. Christianity offers a kind of eschatology *within* an eschatology, whereas Islam uses Jewish and Christian themes to form a composite narrative of the world's apocalyptic transformation.

Having now plotted the basic sequence of events that dispensationalist Christians, Sunnis, and Shias believe will occur in the end, let's turn to how these beliefs are shaping the course of world history.

Notes

1. See Michael Svigel, "A History of Dispensationalism in Seven Eras," in *Dispensationalism and the History of Redemption: A Developing and Diverse Tradition*, ed. Jeffrey Bingham and Glenn Kreider (Moody Publishers, forthcoming). Thanks to Michael Svigel for sending me a draft of this chapter.

2. See John Walvoord, "Reflections on Dispensationalism," July 18, 2007, http://www.walvoord.com/article/151.

3. Again, see John Walvoord, "Reflections on Dispensationalism," July 18, 2007, http://www.walvoord.com/article/151.

4. See "An Evening of Eschatology—John Piper, Wilson, Storms and Hamilton," YouTube video, uploaded October 19, 2011, https://www.youtube.com/watch?v=ws0vbT4Yu2s.

5. See appendix 2 for more.

6. See "What Happens after Death?" Got Questions? http://www.gotquestions.org/what-happens-death.html.

7. As a Compelling Truth article puts it, "The first three—'to finish transgression, to put an end to sin, to atone for wickedness'—were accomplished

with Jesus' sacrifice on the cross, but have not yet been applied to Israel as a people. The last three have not occurred at all, so we know the seventieth 'seven' is yet to come." See "What Is the End Times Tribulation?" Compelling Truth, http://www.compellingtruth.org/end-times-tribulation.html.

8.   See "What Is the Tribulation? How Do We Know the Tribulation Will Last Seven Years?" Got Questions? http://www.gotquestions.org/Tribulation.html.

9.   See "What Is the Tribulation? How Do We Know the Tribulation Will Last Seven Years?" Got Questions? http://www.gotquestions.org/Tribulation.html.

10.   See "Who is the Antichrist," Got Questions? http://www.gotquestions.org/antichrist.html.

11.   See "Are We Living in the End Times?" Got Questions? http://www.gotquestions.org/living-in-the-end-times.html.

12.   See "Who Is the False Prophet of the End Times?" Got Questions? http://www.gotquestions.org/false-prophet.html.

13.   Some early Church Fathers, such as Irenaeus, were aware of this variant.

14.   Similar wording can be found at "What Is the End Times Tribulation?" Compelling Truth, http://www.compellingtruth.org/end-times-tribulation.html.

15.   See "What Is the End Times Tribulation?" Compelling Truth, http://www.compellingtruth.org/end-times-tribulation.html.

16.   See "Who Is the False Prophet of the End Times?" Got Questions? http://www.gotquestions.org/false-prophet.html.

17.   Thanks to Shea Houdmann from Got Questions? for help on this point.

18.   According to the chronology given here: "What Is Going to Happen in the End Times?" Compelling Truth, http://www.compellingtruth.org/end-times.html.

19.   See "What Is the End Times Tribulation?" Compelling Truth, http://www.compellingtruth.org/end-times-tribulation.html. It's worth pointing out here that some dispensationalists interpret the scope of these prophecies as being limited to the land and seas around Israel, such as the Mediterranean Sea, rather than applying to the world as a whole.

20.   See "What Is the Land That God Promised to Israel," Got Questions? http://www.gotquestions.org/Israel-land.html.

21.   See "What Is the Millennial Kingdom, and Should It Be Understood Literally?" Got Questions? http://www.gotquestions.org/millennium.html.

22. We should note here that not all dispensationalists accept a distinct Palestinian Covenant.

23. See "What Is Going to Happen in the End Times?" Compelling Truth, http://www.compellingtruth.org/end-times.html.

24. See "What Is the Millennium/Millennial Kingdom?" Compelling Truth, http://www.compellingtruth.org/millennium.html.

25. See "What Is Going to Happen in the End Times?" Compelling Truth, http://www.compellingtruth.org/end-times.html.

26. See "What Is Going to Happen in the End Times?" Compelling Truth, http://www.compellingtruth.org/end-times.html.

27. See Revelation 20:14.

28. See "What Is Going to Happen in the End Times?" Compelling Truth, http://www.compellingtruth.org/end-times.html.

29. Although, as Marcia Hermansen points out (personal communication), this could have been a waking vision.

30. See John Esposito, "Great World Religions: Islam," *The Great Courses* (Teaching Company, 2003).

31. Again, see John Esposito, "Great World Religions: Islam," *The Great Courses* (Teaching Company, 2003).

32. This is actually far too simplistic. The prohibition isn't limited to Muhammad: it applies also to Jesus and Moses, and even to living things in general. While cartoonists in recent times have been murdered for depicting Muhammad, historically speaking, the prohibition has not been especially strict. Thanks to William Chittick and Marcia Hermansen for pointing this out to me.

33. See John Esposito, *Islam: The Straight Path* (Oxford University Press, 2011), p. 85. See also John Esposito, "Great World Religions: Islam," *The Great Courses* (Teaching Company, 2003).

34. Technically, "Shia" means "party," as in the "party of Ali." The term "Shiite," in contrast, refers to a member of the party. For present purposes, I've decided to simplify and use "Shias" rather than "Shiites" to mean the followers themselves.

35. See Ron Geaves, *Islam Today: An Introduction* (Continuum International Publishing Group, 2010), pp. 23–24.

36. See Mike Shuster, "The Origins of the Shiite-Sunni Split," *NPR*, February 12, 2007, http://www.npr.org/sections/parallels/2007/02/12/7332087/the-origins-of-the-shiite-sunni-split.

37.   See Mike Shuster, "The Origins of the Shiite-Sunni Split," *NPR*, February 12, 2007.

38.   See Jean-Pierre Filiu, *Apocalypse in Islam* (University of California Press, 2011), p. 8.

39.   Again, see Jean-Pierre Filiu, *Apocalypse in Islam* (University of California Press, 2011), p. 8.

40.   See Ron Geaves, *Islam Today: An Introduction* (Continuum International Publishing Group, 2010), p. 26. Geaves also writes that "the Shi'a developed the theological and political position that Ali should by right have been the first Caliph, handing the Caliphate on to the 'People of the Household', the Prophet's direct family descendants, thus giving the leadership of the Muslims over to a rule of the bloodline that mysteriously contained some of Muhammad's spiritual power and authority" (p. 24).

41.   See David Cook, "Messianism in the Shiite Crescent," Hudson Institute, April 8, 2011, http://www.hudson.org/research/7906-messianism-in-the-shiite-crescent.

42.   Thanks to William Chittick for clarifying this point for me.

43.   See William Chittick, "Muslim Eschatology," in *Oxford Handbook of Eschatology*, ed. Jerry Walls (Oxford University Press, 2008), p. 133.

44.   See David Cook, *Contemporary Muslim Apocalyptic Literature* (Syracuse University Press, 2005), p. 7.

45.   See Marcia Hermansen, "Eschatology," in *The Cambridge Companion to Classical Islamic Theology*, ed. Tim Winter (Cambridge University Press, 2008), p. 310.

46.   See Jean-Pierre Filiu, *Apocalypse in Islam* (University of California Press, 2011), p. 6.

47.   I refer here to David Cook (2005), Jean-Pierre Filiu (2011), and Yasir Qadhi (2009). This section is *greatly indebted* to all these sources. Qadhi's lecture series, "The Mahdi Between Fact and Fiction," 2009, can be found on YouTube at https://www.youtube.com/watch?v=1oCf7ae__kk.

48.   The relevant hadith states: "The Prophet said: If only one day of this world remained. Allah would lengthen that day (according to the version of Za'idah), till He raised up in it a man who belongs to me or to my family whose father's name is the same as my father's, who will fill the earth with equity and justice as it has been filled with oppression and tyranny (according to the version of Fitr). Sufyan's version says: The world will not pass away before the Arabs are ruled by a man of

my family whose name will be the same as mine." See also Jane Idleman Smith and Yvonne Yazbeck Haddad, *The Islamic Understanding of Death and Resurrection* (Oxford University Press, 2002), p. 69.

49.   The relevant hadith states: "The Prophet said: The Mahdi will be of my stock, and will have a broad forehead and a prominent nose. He will fill the earth will equity and justice as it was filled with oppression and tyranny, and he will rule for seven years." See Yasir Qadhi, "The Mahdi Between Fact and Fiction," lecture, 2009, https://www.youtube.com/watch?v=1oCf7ae__kk.

50.   See David Cook, *Contemporary Muslim Apocalyptic Literature* (Syracuse University Press, 2005), p. 128.

51.   The relevant hadith states (once again): "The Prophet said: The Mahdi will be of my stock, and will have a broad forehead a prominent nose. He will fill the earth will equity and justice as it was filled with oppression and tyranny, and he will rule for seven years." See Jane Idleman Smith and Yvonne Yazbeck Haddad, *The Islamic Understanding of Death and Resurrection* (Oxford University Press, 2002), p. 70.

52.   See Yasir Qadhi, "The Mahdi Between Fact and Fiction," lecture, 2009, https://www.youtube.com/watch?v=1oCf7ae__kk.

53.   See David Cook, "Messianism in the Shiite Crescent," Hudson Institute, April 8, 2011, http://www.hudson.org/research/7906-messianism-in-the-shiite-crescent. As Cook writes, "At this particular junction, the classical sources say the Mahdi will appear, either in the region of Mecca and Medina (associated with the time of the hajj pilgrimage) or in the region of Khorasan (eastern Iran and Afghanistan). Alternative places associated with the Mahdi are his future capital of Kufa (in southern Iraq) and the messianic pilgrimage site of Jamkaran (near Qom in Iran), where he is traditionally believed to be located or at least accessible."

54.   This comes from a weak hadith, which states, "Disagreement will occur at the death of a caliph and a man of the people of Medina will come flying forth to Mecca. Some of the people of Mecca will come to him, bring him out against his will and swear allegiance to him between the Corner and the Maqam. An expeditionary force will then be sent against him from Syria but will be swallowed up in the desert between Mecca and Medina. When the people see that, the eminent saints of Syria and the best people of Iraq will come to him and swear allegiance to him between the Corner and the Maqam." See Jean-Pierre Filiu, *Apocalypse in Islam* (University of California Press, 2011), p. 21.

55.   The relevant hadith states: "Allah's Messenger, what about him who would be made to accompany this army willy nilly? Thereupon he said: He would be

made to sink along with them but he would be raised on the Day of Resurrection on the basis of his intention."

56. This comes from an often-quoted weak hadith: "While we were with the Messenger of Allah, some youngsters from Banu Hashim came along. When the Prophet saw them, his eyes filled with tears and his color changed. I said: 'We still see something in your face that we do not like (to see)'. He said: 'We are members of a Household for whom Allah has chosen the Hereafter over this world. The people of my Household will face calamity, expulsion and exile after I am gone, until some people will come from the east carrying black banners. They will ask for something good but will not be given it. Then they will fight and will be victorious, then they will be given what they wanted, but they will not accept it and will give leadership to a man from my family. Then they will fill it with justice just as it was filled with injustice. Whoever among you lives to see that, let him go to them even if he has to crawl over snow.'"

57. The relevant hadith states: "There would be in the last phase of the time a caliph who would distribute wealth but would not count."

58. Quoted in Jean-Pierre Filiu, *Apocalypse in Islam* (University of California Press, 2011), p. 14.

59. To quote the relevant hadith: "The Prophet looked out at us from a room, when we were talking about the Hour. He said: 'The Hour will not begin until there are ten signs: Dajjal, (False Christ), the smoke, and the rising of the sun from the west.'"

60. See Yasir Qadhi, "The Mahdi Between Fact and Fiction," lecture, 2009, https://www.youtube.com/watch?v=1oCf7ae__kk.

61. In full, the relevant hadith states: "The Prophet said (about Ad-Dajjal) that he is one eyed, his right eye is as if a protruding out grape."

62. See David Cook, *Contemporary Muslim Apocalyptic Literature* (Syracuse University Press, 2005), p. 9.

63. The relevant hadith, which is strong, states: "We asked: How long will he remain on the earth ? He replied: Forty days, one like a year, one like a month, one like a week, and rest of his days like yours."

64. See Yasir Qadhi, "The Mahdi Between Fact and Fiction," lecture, 2009, https://www.youtube.com/watch?v=1oCf7ae__kk.

65. The relevant hadith, which is strong, states: "Constantinople will be conquered with the coming of the Hour." Another hadith states: "The flourishing state of Jerusalem will be when [Medina] is in ruins, the ruined state of [Medina] will be when the great war comes, the outbreak of the great war will be at the

188 • THE END

conquest of Constantinople and the conquest of Constantinople when the Dajjal (Antichrist) comes forth."

66.   The relevant hadith states: "You have heard about a city, one side of which is on land and the other is in the sea (Constantinople). They said: Allah's Messenger, yes. Thereupon he said: The Last Hour would not come unless seventy thousand persons from Bani Ishaq would attack it. When they would land there, they will neither fight with weapons nor would shower arrows but would only say: 'There is no god but Allah and Allah is the Greatest,' and one side of it would fall. Thaur (one of the narrators) said: I think that he said: The part by the side of the ocean. Then they would say for the second time: 'There is no god but Allah and Allah is the Greatest' and the second side would also fall, and they would say: 'There is no god but Allah and Allah is the Greatest,' and the gates would be opened for them and they would enter therein and, they would be collecting spoils of war and distributing them amongst themselves when a noise would be heard saying: Verily, Dajjal has come. And thus they would leave everything there and go back."

67.   See David Cook, *Contemporary Muslim Apocalyptic Literature* (Syracuse University Press, 2005), p. 133.

68.   See Jean-Pierre Filiu, *Apocalypse in Islam* (University of California Press, 2011), p. 16.

69.   The relevant hadith states: "The Dajjal would be followed by seventy thousand Jews of Isfahan wearing Persian shawls."

70.   The relevant hadith states: "By the One in Whose Hand is my soul! Ibn Mariam [son of Mary, or Jesus] shall soon descend among you, judging justly. He shall break the cross, kill the pig, remove the Jizyah [on non-Muslims], and wealth will be so bountiful that there will be none to accept it."

71.   Quoted in Jean-Pierre Filiu, *Apocalypse in Islam* (University of California Press, 2011), p. 17.

72.   Again, see Jean-Pierre Filiu, *Apocalypse in Islam* (University of California Press, 2011), p. 17.

73.   The relevant hadith states: "Gog and Magog people will be set free and they will emerge as Allah says: 'swoop(ing) down from every mound.' They will spread throughout the earth, and the Muslims will flee from them until the remainder of the Muslims are in their cities and fortresses, taking their flocks with them. They will pass by a river and drink from it, until they leave nothing behind, and the last of them will follow in their footsteps and one of them will say: 'There was once water in this place.' They will prevail over the earth, then their leader will say: 'These are the people of the earth, and we have finished them off. Now let us fight the people of heaven!' Then one of them will throw his spear towards the

sky, and it will come back down smeared with blood. And they will say: 'We have killed the people of heaven.' While they are like that, Allah will send a worm like the worm that is found in the noses of sheep, which will penetrate their necks and they will die like locusts, one on top of another. In the morning the Muslims will not hear any sound from them, and they will say: 'Who will sell his soul for the sake of Allah and see what they are doing?' A man will go down, having prepared himself to be killed by them, and he will find them dead, so he will call out to them: 'Be of good cheer, for your enemy is dead!' Then the people will come out and let their flocks loose, but they will not have anything to graze on except their flesh, and they will become very fat as if they were grazing on the best vegetation they ever found.'"

74. See David Cook, *Contemporary Muslim Apocalyptic Literature* (Syracuse University Press, 2005), p. 10.

75. The relevant hadith states: "The Messenger of Allah said: 'The Kabah will be destroyed by Dhul-Suwaiqatan (one with thin legs) from Ethiopia.'" See Jean-Pierre Filiu, *Apocalypse in Islam* (University of California Press, 2011), p. 15.

76. See Jane Idleman Smith and Yvonne Yazbeck Haddad, *The Islamic Understanding of Death and Resurrection* (Oxford University Press, 2002), p. 71.

77. Again, see Jane Idleman Smith and Yvonne Yazbeck Haddad, *The Islamic Understanding of Death and Resurrection* (Oxford University Press, 2002), p. 71.

78. See Jane Idleman Smith and Yvonne Yazbeck Haddad, *The Islamic Understanding of Death and Resurrection* (Oxford University Press, 2002), pp. 78–79.

79. Incidentally, so does "bin," as in Osama bin Laden, or "Osama, son of Laden."

80. See David Cook, "Messianism in the Shiite Crescent," Hudson Institute, April 8, 2011, http://www.hudson.org/research/7906-messianism-in-the-shiite-crescent.

81. See Jean-Pierre Filiu, *Apocalypse in Islam* (University of California Press, 2011), p. 26.

82. See David Cook, "Messianism in the Shiite Crescent," Hudson Institute, April 8, 2011, http://www.hudson.org/research/7906-messianism-in-the-shiite-crescent. See also Jean-Pierre Filiu, *Apocalypse in Islam* (University of California Press, 2011), p. 27.

83. Again, see David Cook, "Messianism in the Shiite Crescent," Hudson Institute, April 8, 2011, http://www.hudson.org/research/7906-messianism-in-the-shiite-crescent.

84. See Jean-Pierre Filiu, *Apocalypse in Islam* (University of California Press, 2011), p. 27.

85. See Jean-Pierre Filiu, *Apocalypse in Islam* (University of California Press, 2011), p. 25.

86. Again, see Jean-Pierre Filiu, *Apocalypse in Islam* (University of California Press, 2011), p. 25.

87. See David Cook, "Messianism in the Shiite Crescent," Hudson Institute, April 8, 2011, http://www.hudson.org/research/7906-messianism-in-the-shiite-crescent.

88. See Yasir Qadhi, "The Mahdi Between Fact and Fiction," lecture, 2009, https://www.youtube.com/watch?v=1oCf7ae__kk.

89. These paragraphs have relied on Jean-Pierre Filiu, "The Return of Political Mahdism," Hudson Institute, May 21, 2009, http://www.hudson.org/research/9891-the-return-of-political-mahdism, and David Cook, "Messianism in the Shiite Crescent," Hudson Institute, April 8, 2011, http://www.hudson.org/research/7906-messianism-in-the-shiite-crescent.

90. See Jean-Pierre Filiu, *Apocalypse in Islam* (University of California Press, 2011), p. 28.

91. See Jean-Pierre Filiu, *Apocalypse in Islam* (University of California Press, 2011), p. xi.

13

Guns, God, and Armageddon

• • • • •

The Clash of Eschatologies

As the historian Charles Townshend notes in his book *Terrorism*, the nature of terrorism has undergone a metamorphosis since the Iron Curtain fell in the early 1990s. While terrorism for much of the preceding decades was motivated by political issues, "the most important defining characteristic of terrorism today" has become "the religious imperative."[1] In other words, the contemporary terrorist is less a political radical with nationalist objectives than a zealot guided by religious convictions about what the world is like and, more importantly, how it *ought to be* in the future.

To be sure, material conditions *do* figure into the equation of terrorist motivation. It would be far too simplistic to *flatly blame* religion for modern terrorism, as some critics of Islam have done. The Islamic State, for example, specifically cites as a cause of indignation the 1916 Sykes-Picot Agreement, whereby the British and French (with Russian involvement) secretly divided up the Middle East into distinct "spheres of influence." In fact, the Islamic State tweeted pictures in 2014 of bulldozers plowing over earthen barriers separating Iraq and Syria, triumphantly announcing that, in doing this, "they were destroying the 'Sykes-Picot' border."[2] Other Islamic State grievances include US support for monarchs and brutal regimes in the region, and the 2003 preemptive invasion of Iraq, which galvanized large numbers of young men to fight what they saw as an unjust, imperialist force participating in a long tradition of Western interference.

But such factors do not diminish the causal role that religion has had in shaping the decisions and actions of terrorists. As stated in the previous chapter, beliefs about the future affect actions in the present. If this is true on the mundane level—and it clearly is, as anyone who gets a flu shot or has life insurance must agree—then it's certainly even truer when the

191

beliefs concern matters of ultimate spiritual importance. In many cases, religion provides that extra *something* needed for, as the Nobel laureate Steven Weinberg puts it, "good people to do bad things." Put differently, an organization like al-Qaeda isn't a Mensa club for psychopaths. Many terrorists who join such groups are well-educated, with good jobs and stable marriages. They may be angry about American jingoism, but what turns them into martyrs rather than opinion-piece writers for *Alternet* or *Al Jazeera* is their deeply held, dogmatic beliefs about the will of God.[3]

One might argue that of all the religious beliefs, those concerning eschatology are the most powerful, at least with respect to calling large numbers of people to action.* After all, what could be more exciting than participating in the climactic battle between Good and Evil at the apotheosis of cosmic history? What could be more significant than giving one's life for God as this weary world comes to a glorious end? I would argue that the most conspicuous, significant, and worrisome form of terrorism in the future won't just be religious in nature, but *apocalyptic*. There are several reasons for thinking this. First, apocalyptic groups have nothing to lose by escalating conflict to new levels of violence. Heaven awaits the holy warrior—the "Christian soldier" (onward!) and Islamic martyr—if death occurs, and for a new, better, utopian world to be established the old one must be destroyed. As Jessica Stern and J. M. Berger observe in their book *ISIS: The State of Terror*, apocalyptic groups aren't "inhibited by the possibility of offending their political constituents because they see themselves as participating in the ultimate battle." As a result, they are "the most likely terrorist groups to engage in acts of barbarism."[4]

Second, while past millennia have seen innumerable groups with apocalypse fever come and go, the fact is that humanity now has an array of annihilation mechanisms that weren't available, or even imagined, just 70 years ago. Advanced dual-use artifacts—from synthetic genomics and other biotechnologies to future anticipated molecular manufacturing—are making it genuinely possible for groups, and even single individuals

---

* As Richard Landes points out in a panel discussion with Graeme Wood and Will McCants, paraphrasing the scholar Stephen O'Leary in his book **Arguing the Apocalypse** (Oxford University Press, 1994), "Apocalyptic rhetoric [is] the hardest rhetoric to get people to adopt, but once you draw them in it's the most powerful rhetoric imaginable." See "Q&A Richard Landes, William McCants & Graeme Wood—4," YouTube, video, published May 15, 2015, https://www.youtube.com/watch?v=joXLQa8zE2Y.

working in complete isolation, to bring about global catastrophes of unfathomable proportions. Thus, a death cult that *truly believes* that the realization of paradise depends on the decimation of society could, for the first time in human history, actually catalyze such an event. Just as religious terrorism has come to supplant political terrorism, apocalyptic terror will likely emerge as one of the key issues of the coming decades, as the technologies discussed throughout this book reach full maturity.

While no terrorist group today possesses the means for instantiating an existential risk, eschatological thinking constitutes a significant, albeit often invisible, force in world affairs. The result is a phenomenon that I call the *clash of eschatologies* (on the model of Samuel Huntington's "clash of civilizations," although the clash of eschatologies is only tangentially related to Huntington's idea). This refers to the antagonistic interaction between mutually incompatible belief systems specifically focused on how the world, through human and supernatural means, will come to an end.

Many of the most significant and defining struggles of our age, I would argue, have their origin in this clash and are being indefinitely perpetuated by it. I'm not the only observer to make this claim. Philosopher of religion Jerry Walls, for example, writes in the *Oxford Handbook of Eschatology* that "it is not too much to say that some of the most passionately contested cultural, political, and social conflicts in our world today are rooted in *competing eschatological claims*."[5] Similarly, French historian Jean-Pierre Filiu, writing in his award-winning 2012 book *Apocalypse in Islam*, describes the situation of prophecy-minded Christians in the US and apocalypse-oriented actors in the Middle East as "therefore less a clash of civilizations that is now beginning to take shape than a *confrontation of millenarianisms*."[6] And Australian philosopher Matthew Sharpe, in a paper on the "return to religion," refers to, as he puts it, the "post-2001 *clash of fundamentalisms*, to be distinguished from Huntington's famous 'clash of civilizations.'"[7]

This fundamentalist, millenarian, eschatological showdown is a genuinely important phenomenon in the world, and one that must not be ignored by riskologists. To grasp the full reality of our situation, the secular eschatologist must understand not only the dangers posed by advanced dual-use technologies, anthropogenic phenomena like global warming, and natural events such as supereruptions but also how the fatalistic end-times convictions of Christians and Muslims (not to mention the many smaller religious cults that pop up here and there) are being actively *applied* in the real world. The following two sections offer a thumbnail sketch of this situation, beginning with religious leaders in the US and then turning to the terrorists of modern Mesopotamia. I should add that nothing about this chapter's theses depend on the particulars examined below; the central idea is that millenarian thinking is something like a human universal, and as such it will persist long after Pastor John Hagee and Abu Bakr al-Baghdadi are gone. (See appendix 3 for some examples of millenarian thinking in a "secular" context.)

What Would Jesus Do?

Zionism is the nationalist-political view that there should exist a Jewish state in the sliver of Middle Eastern territory called Palestine. While Zionism began as a largely secular, Jewish response to growing anti-Semitism in Europe during the nineteenth century, it later spawned a strand of pro-Israeli sentiment in the West called *Christian Zionism*. This movement was rooted not in a desire for the Jews to escape persecution but in a longing for the eschatological narrative of dispensationalism to come true. Thus, as the sociologist and pastor Tony Campolo notes, "without understanding dispensationalism . . . it is almost impossible to understand how Christian Zionism has come to dominate American Evangelicalism and been so influential on the course of US Middle East policy."[8] Israel matters to the dispensationalists because, as the previous chapter revealed, *everything about the end times revolves around the existence of a Jewish state in Palestine.* (Recall that the land promised to the Jews includes not just all of Israel but also the entire territories of Palestine and Jordan, plus some parts of Egypt, Syria, Saudi Arabia, and Iraq.)

John Nelson Darby himself was an advocate of creating a Jewish state in the region, but it was his friend William Blackstone (1841–1935) who was quite possibly the most vocal early champion of Christian Zionism. According to Blackstone, the immigration of Jews to Palestine in the late nineteenth century (before a Jewish state was established) was a clear "sign of the times," and Blackstone held that "the United States [has] a special role and mission in God's plans for humanity: that of a modern Cyrus, to help restore the Jews to Zion."* This view was, more or less, shared by

* Cyrus the Great was the leader who founded the Achaemenid Empire. Upon conquering the Babylonian empire, he freed the Jews from captivity and allowed them to resettle in Jerusalem. See Yaakov Ariel, "An Unexpected Alliance: Christian Zionism and Its Historical Significance," **Modern Judaism** 26:1 (2006).

Arthur Balfour, the foreign secretary of the UK from 1916 to 1919. Balfour is perhaps most well known for penning the Balfour Declaration (1917), which made explicit the UK's support for the creation of a Jewish state. As the declaration puts it: "His Majesty's government view with favour the establishment in Palestine of a national home for the Jewish people, and will use their best endeavours to facilitate the achievement of this object." Balfour grew up in a dispensationalist church and was "publicly committed to the Zionist agenda for [both] 'biblical' and colonialist reasons."[9] Consequently, as Sam Harris notes in *The End of Faith*, the Balfour Declaration "was inspired, at least in part, by a conscious conformity to biblical prophecy."[10]

The establishment of a Jewish state became a reality in 1948, three years after Hitler's genocidal empire was defeated by the Allied powers. From the dispensationalist perspective, this was the most significant eschatological event since the Jewish temple was destroyed in 70 AD (after which the last Jewish diaspora commenced). One can find quotes from just about every major dispensationalist leader affirming, often with great vehemence and excitement, the importance of 1948. In the eyes of many Christians, it was unequivocal evidence that biblical prophecy is true—and therefore that Armageddon is quickly approaching.

Further prophetic confirmation came in June 1967 after a build-up of Egyptian forces led Israel to preemptively strike a number of surrounding nations in what is known as the Six-Day War, severely compromising their military capabilities. In the process, Israel expanded its borders, taking control of the holy city of Jerusalem, from which Jesus is said to reign during the Millennium. As the scholar Stephen Spector notes in *Evangelicals and Israel*, "For many devout Christians, the Israeli conquest of the old city of Jerusalem in 1967 also was a crucial event, [one] that Jesus himself foretold" in Luke 21:24, where Jesus declares that "they will fall by the edge of the sword, and be led captive among all nations; and Jerusalem will be trodden down by the Gentiles, until the times of the Gentiles are fulfilled."[11] Thus, with Jerusalem under Jewish control, dispensationalists saw the stage nearly set for the climactic battle between Jesus and the Antichrist in the holy lands.

For most of the twentieth century, Darby's invented theology remained relatively obscure, although it was influential among a segment of the Christian community. This changed dramatically in the US with the publication of Hal Lindsey's mega-selling book *The Late Great Planet Earth*

196 • THE END

in 1970. It was, in fact, the best-selling nonfiction book of that decade (using the term "nonfiction" loosely, as Bart Ehrman quips[12]). Lindsey's approach was quintessentially dispensationalist, and his message was urgent: the end is near; prepare for the Rapture or be left behind to suffer through the Tribulation. Lindsay's success as a popularizer of dispensationalist eschatology opened up the door for similarly themed books to hit the best-seller list, such as Tim LaHaye and Jerry Jenkins' *Left Behind* series, which has sold over 700 million copies worldwide. "To put this into perspective," Campolo writes, "[these] books have sold more copies than the complete works of Stephen King and John Grisham combined."[13]

The "enormous appeal" of this eschatology, made accessible in books and films, resulted in it becoming arguably the most influential eschatology of the late twentieth century, especially among nonacademic Christians.* Consequently, as Michael Sells writes in the *Oxford Handbook of Religion and Violence*, since the late 1970s, "dispensationalism has dominated the Christian Zionist movement in the United States."[14] The popularity of such ideologies among the American public explains, at least in part, why US presidents of all political (and religious) persuasions have generally supported Israel as an article of faith: doing the opposite would outrage tens of millions of people with the power to vote. To date, US military aid for Israel has totaled roughly $233.7 billion. One of President Obama's campaign promises was, in fact, to "implement a Memorandum of Understanding that provides $30 billion in assistance to Israel over the next decade"—a commitment that his administration has kept, despite occasional flare-ups with Israeli prime minister Benjamin Netanyahu.[15]

But voters aren't the only ones under the spell of dispensationalism; a large number of political leaders have either espoused dispensationalist beliefs or closely allied themselves with the dispensationalist community. While Ronald Reagan apparently shied away from his Armageddon-eager

* As Jerry Walls writes, "While elite academic theologians might find the enormous appeal of fundamentalist eschatology somewhat amusing, if not embarrassing or even annoying, they can hardly afford to ignore it, as much as they might prefer to do so. Consider the following comment by Harold Attridge, dean of Yale Divinity School: 'Much of this eschatological fascination could be dismissed as misguided nonsense, but for the fact that so many people take it seriously, allowing it to form their image of a Rambo-like Christ, whose future violence against the powers of evil we might emulate now.'" See Jerry Walls, "Introduction," in the **Oxford Handbook of Eschatology**, ed. Jerry Walls (Oxford University Press, 2008), p. 10.

views toward the second half of his presidency, he came to power as a fervent dispensationalist, and his beliefs influenced the policies he pursued.[16] During a 1971 dinner, for example, he said the following:

> Everything is falling into place. It can't be too long now. Ezekiel says that fire and brimstone will be rained upon the enemies of God's people. That must mean that they'll destroyed by nuclear weapons. They exist now, and they never did in the past. Ezekiel tells us that Gog, the nation that will lead all of the other powers of darkness against Israel, will come out of the north. Biblical scholars have been saying for generations that Gog must be Russia. What other powerful nation is to the north of Israel? None.[17]

This wasn't the end of it. The individual that Reagan appointed to be the secretary of the interior, James Watt, was an ardent dispensationalist who "believed that the world would end before the oil was used up and before we suffered the consequences of global warming or deforestation, so it was almost our duty to be profligate with Earth's divinely provided resources." Reagan also included "men like Jerry Falwell and Hal Lindsey in his national security briefings," as Harris notes.[18] Falwell was so powerful at the time that after Israel bombed Iraq in 1981, he was the first person— even before Reagan—that the then Israeli prime minister Menachem Begin called, asking Falwell to "explain to the Christian public the reasons for the bombing."* Thus, whatever Reagan's private beliefs at the beginning and end of his two terms, he surrounded himself with dispensationalists who wouldn't have minded a warm end to the Cold War.

Some scholars have dubbed the clan of dispensationalist leaders with political power in Washington "the Armageddon lobby." From this lobby's perspective, *bad news in the world, especially in the Middle East, is decidedly good news.* The reason is that, as Walls puts it, "dispensationalist eschatology inclines its adherents not only to despair of changing the world for good, but even to take a certain grim satisfaction in the face of wars and natural disasters, events which they interpret as the fulfillment of prophecy pointing to the end of the world."[19] For such believers, a mushroom cloud rising up

---

* Similarly, "when Prime Minister Benjamin Netanyahu visited the United States in 1998, he called first on Falwell, and only then met with President Clinton." See Alan Mittleman et al., **Uneasy Allies?: Evangelical and Jewish Relations** (Lexington Books, 2007), p. 55.

over Mesopotamia would be an occasion not for horror, but eschatological elation, since great wars and devastation must occur before our eyes will see "the glory of the coming of the Lord" (to quote "The Battle Hymn of the Republic"). A shocking example of such ghoulish delight occurred in 2006, when war broke out between Israel and Hezbollah (a Lebanon-based Shia Islamist militant group). The nationally syndicated radio show host and Christian Zionist, Janet Parshall, proclaimed "in a voice brimming with joy" that "these are the times we've been waiting for. . . . This is straight out of a Sunday school lesson"![20]

Some of the most powerful members of the Armageddon lobby take the pessimism of their premillennial eschatology one step further and attempt to actually *accelerate* the end times by actively fomenting the conditions necessary for it to occur. This is applied eschatology in action, and with real-world consequences. As the thoroughly nondispensationalist National Council of Churches lamented in 2007, "The Christian Zionist programme provides a worldview where the Gospel is identified with the ideology of empire, colonialism and militarism." By placing "an emphasis on apocalyptic events leading to the end of history," it *actively impedes* "justice and peace in the Middle East." The Jews are meanwhile dehumanized as being "mere pawns in an eschatological scheme," and interfaith dialogue between Jews, Christians, and Muslims is stymied, "since [dispensationalism] views the world in starkly dichotomous terms."[21]

Historically speaking, the Armageddon lobby may have played a role in the lead-up to the 1990 Gulf War, which a shocking 15% of the American population believed was the beginning of Armageddon. At the time, as Philip Lamy notes, American bookstores were overflowing with "books about prophecy and the end of the world," and in fact Lindsey's *The Late Great Planet Earth* saw an 83% spike in sales.[22] The Armageddon lobby also appears to have influenced the 2003 preemptive invasion of Iraq, which Bush told a Palestinian delegation in 2003 he pursued because he was "driven with a mission from God." In Bush's words: "God would tell me, 'George, go and fight these terrorists in Afghanistan.' And I did. And then God would tell me 'George, go and end the tyranny in Iraq.' And I did."*

* Bush apparently said something similar to French president Jacques Chirac, namely: "Gog and Magog are at work in the Middle East. . . . The biblical prophecies are being fulfilled. . . . This confrontation is willed by God, who wants to use this conflict to erase his people's enemies before a New Age begins."

It's unclear the extent to which Bush subscribes (or subscribed) to dispensationalism per se, but the former House majority leader, Dick Armey, claimed in a BBC interview that Bush believes in the Tribulation and "end times." (He then denied that such beliefs influenced Bush's policies.) Either way, Glenn Shuck notes in a chapter for *The Oxford Handbook of Millennialism* that "Bush helped widen Dispensationalists' involvement in the world, even if [his] personal beliefs on the Endtime remain unknown with precision."[23] For example, in 2006, Bush set up a number of "off-the-record" meetings with Christians United for Israel (CUFI), a group that "claims more than [two] million members and has done more than just about any other organization to make Israel a defining foreign-policy issue for evangelical Christians in the United States."[24] *Foreign Policy* magazine describes it as the largest and most powerful pro-Israel group in the US, even larger than the American Israel Public Affairs Committee (AIPAC)—itself one of the biggest and most influential lobbies *in the US*.[25]

CUFI was founded in 2006 by the Texas mega-church pastor and eschatological enthusiast John Hagee, a man who once speculated that "God sent Adolf Hitler to help Jews reach the promised land."[26] Since its inception, CUFI has fostered relationships with a vast range of right-wing politicians and pundits, including (but not limited to) Joseph Lieberman, John McCain, Roy Blunt, John Cornyn, Tom Cotton, Trent Franks, Lindsey Graham, Tim Scott, Sanford D. Bishop, Tom Price, Peter Roskam, Tom DeLay, Glenn Beck, Michele Bachmann, Sarah Palin, Mike Huckabee, Newt Gingrich, Rick Santorum, Eric Cantor, Ted Cruz, and Alan Keyes (the man who lost a 2004 Senate election to a young politician named Barack Obama). As this list suggests, the impact of CUFI on the American political landscape has not been insignificant. Many conservative leaders are closely tied to the organization, and numerous prominent Republicans have spoken at CUFI events. During one such gathering, Lieberman praised Hagee as "a *Ish Elohim*, a man of God and those words really fit him. . . . Like Moses he's becoming a leader of a mighty multitude, even greater than the multitude that Moses led from Egypt to the promised land." And Cruz, a 2016 presidential candidate, recently described joining Hagee on stage at his church as "an honor and a privilege."

The same year that Hagee founded CUFI, he published a book called *Jerusalem Countdown: A Warning to the World*. In it, Hagee superimposes the narrative of dispensationalism onto contemporary world affairs, retelling the story in terms of Russia, Islamic countries, and the European

Union, the latter of which he believes will propel the Antichrist to power. (Incidentally, LaHaye believes the Antichrist will gain power of the United Nations. Both interpretations foment conservative disdain for these institutions.)

The most eschatologically significant entity in Hagee's current worldview, though, is the Islamic Republic of Iran, which he colorfully describes as "the head of the beast of radical Islam in the Middle East."[27] Hagee's fixation on Iran stems from his worry that Iran's nuclear ambitions pose an existential threat to Israel—a worry that might not be unfounded, as we'll see below. Since the grand narrative of history dictated by God in the Bible depends entirely upon Israel's continued existence, it follows that Iran must be stopped from creating a nuclear weapon *at all costs*. This has led Hagee to repeatedly call for the US to join Israel in a *nuclear first strike* against Iran. As Hagee writes, rather chillingly given his political reach, "We are standing on the brink of a nuclear Armageddon. The coming nuclear showdown with Iran is a certainty. Israel and America must confront Iran's nuclear ability and willingness to destroy Israel with nuclear weapons. For Israel to wait is to risk committing national suicide," and therefore to risk disrupting the dispensationalist narrative.[28] By God, this cannot be allowed to happen.

Calls for a preemptive war with Iran have been reverberating through the echo chamber of religious conservatism for years. In a 2014 interview with the *Washington Free Beacon*, Michele Bachmann said that her last words to Obama before leaving Congress were "Mr. President, you need to bomb the Iranian nuclear facilities, because if you don't, Iran will have a nuclear weapon on your watch and the course of world history will change."[29] In the context of dispensationalism, "the course of world history changing" takes on a quite different meaning than it would otherwise have. The following year, in fact, Bachmann gave a series of interviews in which she said that "we need to realize how close this clock is to getting towards the midnight hour," referring to the end times. "The good news that I want to transition to," she continued, "is that, remember, the prophets said in the Old Testament, they longed to look into the days that we live in, they long to be part of these days. . . . We can talk about God's time clock and the fact that Jesus Christ's return is imminent. Is there anything more important?"[30]

No sane person wants Israel erased from the map in a nuclear blaze, and there may be good reasons for fearing Iran and its nuclear ambitions. But the motivations driving Bachmann, Hagee, and their dispensationalist colleagues to support Israel are duplicitous: avoiding a nuclear holocaust isn't of humanitarian, strategic, or even oil-related importance; it's of *biblical importance*. As Hagee exclaimed at a CUFI event in 2007, "Let us shout it from the housetops that a new day has been born in America. The sleeping giant of Christian Zionism has awakened. If a line has to be drawn, draw the line around Christians and Jews. We are united. We are indivisible. And together we can reshape history."[31]

Despite such rhetoric of unity, many Jews are uneasy allying themselves with dispensationalists. After all, according to the dispensationalist narrative, their ultimate fate will involve either converting to Christianity or being slaughtered during the Tribulation and eventually sent to hell. While Christian Zionists motivated by dispensationalism appear to be, superficially, on the same side as the Jews, their beliefs could hardly be more insidiously, even genocidally, anti-Semitic. Nonetheless, given the plight of Jews in the Middle East, surrounded by hostile states, many Jews are willing take help where they can find it. The uncomfortable differences in long-term goals can be overlooked if the short-term goals are sufficiently well-aligned.

Don't Shoot the Messenger

Many adherents of Islam, it turns out, are no less infatuated with eschatological fantasies about the imminent end of history than Christian apocalypticists. The Gestalt shift here can be quite striking: peering at the exact same events, Muslims attribute a completely different meaning and significance to them than Christians. For example, recall that 15% of Americans interpreted the 1990 Gulf War as the beginning of Armageddon; at the same time, the leader of the Nation of Islam, Louis Farrakhan, boldly described it as "that which the scriptures refer to as the War of Armageddon." When the Gulf War turned out not to be Armageddon, Farrakhan claimed it was merely the *precursor* to Armageddon.

Similarly, while "millions of Americans" saw the 2003 invasion of Iraq as "part of an unfolding divine plan," to quote the historian Paul Boyer, many Muslims in the region took it to confirm their own prophetic narratives.[32] A Shia fighter in Baghdad, for example, confessed to Reuters in 2014 that "he knew he was living in the era of the Mahdi's return when the United States and Britain invaded Iraq." The US-led incursion was, as he put it, "the first sign and then everything else followed."[33] And whereas dispensationalists like Hagee and LaHaye see the European Union and

United Nations, respectively, as the "revived Rome" mentioned in Bible prophecy, apocalypticists in the Islamic world tend to identify Rome as the United States, whose soldiers are "crusaders." Thus, looking at the same Wittgensteinian figure, Christians see a duck while Muslims see a rabbit.

The wealthiest and most powerful terrorist organization in history is the Islamic State. According to then US secretary of defense Chuck Hagel, this group of Sunni extremists is "beyond anything that we've seen," with unprecedented capabilities, organization, and funding. It's also motivated by "an apocalyptic, end-of-days strategic vision," as General Martin Dempsey noted in the same press conference with Hagel.[34] The Islamic State's rise to power is a tangle of infighting and rebranding but for background purposes is worth recapitulating in brief. It emerged from al-Qaeda in Iraq (AQI), which took its name shortly after the 2003 US invasion of Iraq and was run by the "merciless" Abu Musab al-Zarqawi. At the time—and up to this writing—al-Qaeda itself was headed by a surgeon named Ayman al-Zawahiri, who took over from bin Laden but even before this was described as the "real brains of the outfit."[35]

In 2005, al-Zawahiri (of al-Qaeda) suggested to al-Zarqawi (of AQI) that AQI should declare an *Islamic state* to "fill security vacuums" left behind when US troops withdraw from the region. Before a state was declared, though, the US Air Force identified al-Zarqawi's location one afternoon, in 2006, and dropped a 500-pound bomb on him. He was replaced as the head of AQI by Abu Ayyub al-Masri, an apocalypticist who had attended an al-Qaeda training camp in 1999 and subsequently became "a confidant of Ayman al-Zawahiri."[36] It was during al-Masri's time as leader that al-Zawahiri's idea was implemented and AQI (along with five smaller groups) was rebranded as the Islamic State of Iraq (ISI). In 2010, al-Masri was killed near the Iraqi city of Tikrit and replaced by a militant jihadist once captured, imprisoned, and then released by the US (during Bush's presidency) called Abu Bakr al-Baghdadi. This is a name to remember, because al-Baghdadi has become quite possibly the most dangerous terrorist in the world today.

While ISI was based in Iraq, another al-Qaeda affiliate named the al-Nusra Front was fighting in Syria, a country whose stability began to break down after the Arab Spring turned ugly.* In 2013, al-Baghdadi announced

---

* It's worth noting that the propaganda arm of the al-Nusra Front is called the White Minaret, for the same eschatological reason that the Islamic State's magazine is named **Dabiq**. Recall from the previous chapter that Jesus' return is prophesied to occur over the white minaret of east Damascus.

that the al-Nusra Front was now under his control. The al-Nusra Front objected to this merger, as did al-Zawahiri. In fact, al-Zawahiri wrote a letter in June 2013 to both the al-Nusra Front and ISI stating his opposition: ISI and the al-Nusra Front should remain separate entities under al-Qaeda's aegis. Al-Baghdadi subsequently replied in an audio message that, despite al-Zawahiri's objections, ISI *will* expand into Syria to become the Islamic State of Iraq and Syria (ISIS). This expansion, along with the extreme brutality of ISIS's strategy, led al-Qaeda to sever its ties with the organization. As its general command put it in an official statement about the controversy, al-Qaeda no longer has "an organizational relationship with [ISIS] and is not the group responsible for their actions."[37]

Almost exactly one year later, ISIS declared that it was establishing a caliphate—the first since the Ottoman Empire (although ISIS doesn't recognize the Ottoman Empire as having had that status). Al-Baghdadi was appointed the caliph, and the organization's name was changed once more from ISIS to simply *the Islamic State* (IS). As of this writing, tens of thousands of foreign recruits from countries like the US, Germany, the Netherlands, Indonesia, Australia, Belgium, France, and the UK are flocking to join the Islamic State.[38] In fact, a 2014 poll found that 16% of French citizens support ISIS, and a survey from July 2015 reports that nearly 1.5 million people in Britain view the Islamic State favorably.[39] The organization is also training children, or "Cubs of the Caliphate," to fight the foreign crusaders, which suggests that the battle against IS-style extremism will be a multigenerational affair.

Just as supporters from around the world travel to the Islamic State's territories, so too has the organization's global reach "snowballed" outward to over 90 countries,[40] including Iraq, Syria, Yemen, Afghanistan, Libya, and Nigeria (through the addition of Boko Haram as an affiliate).[41] As one militant declared in an interview with *Vice News*, the Islamic State hopes to one day expand into the US and, upon doing so, to "raise the flag of Allah in the White House."

As mentioned earlier, the Islamic State has many "material" grievances, including the Sykes-Picot Agreement, the long history of Western support for brutal dictators in the region, the Jewish occupation of Palestine, the 2003 Iraq War, and so on. Some of these may be legitimate complaints of injustice that morally sensitive people in the West can empathize with. In another *Vice News* interview, for example, an Islamic State fighter says, "God willing the Caliphate has been established, and we are going to invade you as you invaded us. We will capture your women as you captured our women. We will orphan your children . . ." at which point he begins to choke up, ". . . as you orphaned our children." He then starts to cry.[42]

But the Islamic State is also animated by an end-times narrative in which they see themselves as active participants. In his much-discussed article for *The Atlantic*, Graeme Wood notes that "during the last years of the US occupation of Iraq, the Islamic State's immediate founding fathers . . . saw signs of the end times everywhere. They were anticipating, within a year, the arrival of the Mahdi." Today, Wood continues, "virtually every major decision and law promulgated by the Islamic State adheres to what it calls, in its press and pronouncements, and on its billboards, license plates, stationery, and coins, 'the Prophetic methodology.'"[43] In another article for the *New Republic*, Wood notes that clerics "on the side of the Islamic State" see al-Baghdadi as the eighth of twelve caliphs who'll rule until the end of history, at which point Jesus will descend in east Damascus to defeat the Dajjal. This is set to happen in 2076, or the year 1500 in the Islamic calendar. The final caliph's name is also already known: it will be Muhammad ibn Abdullah.[44]

The eschatological expectations of the Islamic State account for why its slickly designed online magazine is called *Dabiq*. "As for the name of the magazine," the first issue states, "[Dabiq] was mentioned in a hadith describing some of the events of the Malahim (what is sometimes referred to as Armageddon in English). One of the greatest battles between the Muslims and the crusaders will take place near Dabiq." Indeed, each issue of *Dabiq* has opened with the following quote from AQI's al-Zarqawi: "The spark has been lit here in Iraq, and its heat will continue to intensify— by Allah's permission—until it burns the crusader armies in Dabiq." The editors of the magazine call themselves the "Dabiq team."[45]

In fact, the Islamic State went to great lengths to take control of Dabiq, which has no appreciable strategic or military significance. It's "basically all farmland," as an Islamic State member tweeted, adding that one "could imagine large battles taking place there."[46] The apparent goal of the Islamic State is to lure coalition forces to the city, engage in Armageddon, and thus knock over the first domino of the apocalypse. This is why several beheading videos have taken place in and around Dabiq, including one in which 18 Syrian soldiers are brutally executed. The more inflammatory, the more likely a response. This logic was proven correct after the Islamic State burned alive Jordanian pilot Lt. Muath al-Kasasbeh, an act of unfathomable barbarity that resulted in retaliatory bombings by the Jordanian Air Force. (Interestingly, despite widespread outrage over al-Kasasbeh's death, many Jordanians continue to support the aims of IS.[47])

In bombing the Islamic State, Jordan joined a list of more than *60 countries* currently at war with the group, including Australia, Great Britain, France, Germany, Canada, Saudi Arabia, Egypt, Qatar, Iran, and Turkey. This is three-quarters the number of "banners" prophesied by a hadith to descend upon the Muslims.* Turkey's vow to fight the Islamic State in September of 2014 was especially significant because it made possible "the foreign invasion of northern Syria, meaning from the plain of Dabiq," as one Islamic State fighter posted on Twitter. This added even more fuel to the idea that "the battles (of the End Times) have grown near."[48]

For apocalypticists in the Islamic State—just as for the dispensationalists in John Hagee's church—there will be no peace until the climactic tragedies at the terminus of world history unfold in all their ghoulish horror.[49] As one issue of *Dabiq* avers, "It is clear then that salam (peace) is not the basis of the word Islam . . . the sword will continue to be drawn, raised, and swung until [Jesus] kills the Dajjal (the Antichrist)." Thus, if the true believer wishes for the cessation of violence, the principle of applied eschatology demands that he foment more of it.

One way of accomplishing this goal is, of course, through the exploitation of advanced, dual-use technologies. When the Islamic State overtook the University of Mosul, it confiscated 88 pounds of uranium. Experts say that this material probably can't be used for a dirty bomb (much less for a proper nuclear device), yet one Islamic State member excitedly tweeted that the "Islamic State does have a dirty bomb. We found some radioactive material from Mosul University," adding that "we'll find out what dirty bombs are and what they do. We'll also discuss what might happen if one actually went off in a public area." Later, in early May 2015, the Islamic State suggested in a *Dabiq* article that, with "billions of

* To quote Will McCants, "Jihadi tweets about Dabiq spiked again last month [September 2014] when the United States began to consider military action against the Islamic State in Syria. Islamic State supporters counted the number of nations who had signed up for the 'Rome's' coalition against the Islamic State. 'Thirty states remain to complete the number of eighty flags that will gather in Dabiq and begin the battle.'" See William McCants, "ISIS Fantasies of an Apocalyptic Showdown in Northern Syria," Brookings Institute, October 3, 2014, http://www.brookings.edu/blogs/markaz/posts/2014/10/03-isis-apocalyptic-showdown-syria-mccants.

dollars in the bank," it could purchase a nuclear weapon within a year, possibly from Pakistan (which currently ranks number 22 out of 25 in the Nuclear Threat Initiative's "nuclear materials security index").[50] The article continues: "They [i.e., members of the Islamic State] call on their [official] in Pakistan to purchase a nuclear device through weapons dealers with links to corrupt officials in the region." While smuggling this device into the US through the shorelines and "porous borders" of South and Central America may sound "far-fetched," the article notes that it's "infinitely more possible today than it was just one year ago."

The Islamic State has also expressed interest in advanced biological weaponry. In the summer of 2014, a laptop was confiscated from an Islamic State member with an education in physics and chemistry. Officials found a number of documents about how to weaponize the bubonic plague. As the computer's owner puts it, "The advantages of biological weapons is the low cost and high rate of casualties." He goes on to say that "there are many methods to spread the biological or chemical agents in a way to impact the biggest number of people. Air, main water supplies, food. The most dangerous is through the air." Suicide missions in cars, contaminating air-conditioning systems, and rockets or missiles are also identified as means for inflicting harm.[51] (A less high-tech approach, discussed in a late 2014 *Forbes* article, involves Islamic State members intentionally infecting themselves with Ebola in an effort to spread the disease to other parts of the world.[52]) It's an exercise in nightmares to imagine what such an organization—either the Islamic State or a subsequent group with similar ambitions—would be capable of in just a few decades, as biohacking techniques become more widespread, synthetic biology is increasingly de-skilled, the genomes of new pathogens are published online, 3D gun printing gets more sophisticated, and future artifacts like the nanofactory begin to peek over the horizon of technological feasibility.

At this point, one doesn't need 20/20 vision to see the clash of eschatologies coming into view. Futurological delusions in the world's biggest religions are, and have been for decades, colliding in the single geographic region where all three Abrahamic religions overlap. But the confrontation of millenarianisms is even more complicated than this: it also involves a member of Bush's ill-conceived "Axis of Evil," namely Iran. In fact, this theocratic state[53]—the outcome of Iranians overthrowing an oppressive, corrupt, and ruthless monarch the US itself established and supported (in the place of a democratic government)—had its origins in eschatological expectations. To quote Cook at length:

> It is commonly believed that the Iranian revolution itself was an apocalyptic occurrence, happening as it did in the year 1400 [of the Islamic calendar], and Khomeini [who led the overthrow] skillfully used messianic passions to mobilize ordinary Iranians against the Shah. He framed the revolution, for example, as a struggle against the satanic forces of Yazid (the Shah) by the righteous forces of [Hussein] (the Iranian revolutionaries). He thus recreated in the minds of many the Battle of Karbala as it should have been (with the righteous side winning this time) at the end of the world—a messianic trope from the classical materials.[54]

In the decades since 1979, Iran has had a number of leaders animated by end-times beliefs, perhaps most notably the president of Iran from 2005 to 2013, Mahmoud Ahmadinejad. In 2005, the *Washington Post* quoted him as saying in a speech to Friday Prayer leaders from across Iran, "Our revolution's main mission is to pave the way for the reappearance of the 12th Imam, the Mahdi. Therefore, Iran should become a powerful, developed and model Islamic society."[55] Later, he claimed that, because "of the Mahdi's support for his international message," he found himself "suffused with a halo of light during his speech before the General Assembly of the United Nations."[56] Consequently, "all of a sudden the atmosphere changed there, and for 27–28 minutes all the leaders [literally] did not blink."[57]

In 2012, Ahmadinejad explicitly mentioned the Mahdi in a speech to the UN General Assembly, saying that this messianic figure "will come in the company of Jesus Christ and the righteous," so "let us join hands and clear the way for his eventual arrival with empathy and cooperation, in harmony and unity." On the surface, this speech was quite peaceable. It explicitly emphasized that "the arrival of the Ultimate Savior [the Mahdi], Jesus Christ and the Righteous will bring about an eternally bright future for mankind, not by force or waging wars but through thought awakening and developing kindness in everyone."*

---

* Ahmadinejad expressed a similar sentiment several years earlier: "What is being said about an apocalyptic war and—global war, things of that nature. This is what the Zionists are claiming. . . . The stories that have been disseminated around the world about extensive war, apocalyptic wars, so on and so forth, these are false." See "Transcript: Interview with Iran's Ahmadinejad," NBC, September 18, 2009, http://www.nbcnews.com/id/32913296/ns/world_news-mideastn_africa/print/1/displaymode/1098/.

Yet many of Iran's leaders—from Khomeini in 1979 to the current supreme leader, Ali Khamenei—have described the US in apocalyptic terms as "the Great Satan." Even more ominously, Iran has openly wished for Israel to "vanish from the page of time." As Ahmadinejad put it in 2012, "the very existence of the Zionist regime is an insult to humankind and an affront to all world nations," from which he concluded that "confronting Zionists will also pave the way for saving the whole humankind from exploitation, depravity and misery."[58] This sentiment was echoed by Khamenei in September 2015, shortly after a nuclear deal between Iran and the US was forged by the Obama administration, when he declared, "I'd say (to Israel) that they will not see (the end) of these 25 years. . . . God willing, there will be no such thing as a Zionist regime in 25 years. Until then, struggling, heroic and jihadi morale will leave no moment of serenity for Zionists."[59]

While Ahmadinejad's presidential successor, Hassan Rouhani, holds more moderate religious and political views (being much more of a pragmatist than Ahmadinejad), he apparently attributed his 2013 presidential victory to the Mahdi.[60] Furthermore, the Iranian-American scholar Mehdi Khalaji reports in a paper that "certain factions inside the Revolutionary Guards, consisting mostly of mid-ranking commanders, hold apocalyptic visions." Khalaji explains that such individuals "consider themselves 'soldiers of the Mahdi,'" and "believe a true Shiite cannot merely await the Mahdi without actively engaging in a series of measures to prepare his return."[61] Once again, then, we find the darker shadows of applied eschatology flickering through the mist of religious dogmatism.

It's unclear what exactly Iran's nuclear ambitions are, even after the recent US-Iran nuclear deal. If Iran were to acquire such weapons of mass destruction, the political dynamics of the Middle East would change dramatically, and Israel could face annihilation (a point on which, once again, we can agree with fanatics like Hagee and Bachmann). The cycle here is unfortunate, and potentially catastrophic: Iran sees the US as an existential threat—and why wouldn't it, given its membership in the Axis of Evil and the United States' preemptive invasion of Iraq?* In a letter signed

---

* It's worth recalling here a global survey from 2013 which found that people the world over—from Australia to South Africa, Argentina to Mexico, Spain to Sweden, Russia to Indonesia, Peru to Turkey, Algeria to Germany—see the United States as the number one threat to world peace, with Pakistan more than 16% behind in second.

by "Tehran's lead envoy to the International Atomic Energy Agency," Iran stated that "as long as such threats of military action persist, [it] has no option but (to) protect its security through all means possible, including protection of information which can facilitate openly stated and aggressive military objectives of the warmongers."[62] Meanwhile, many in the US see Iran as an existential threat to Israel—and it may very well be, given bellicose, anti-Semitic statements like those quoted above. This belief has led to calls among US politicians, many of them with dispensationalist ties, for a possible "all options on the table" military intervention in Iran, as discussed in the previous section.

But Iran's influence in the region since 2003 has become far more insidious. Many radical Shia militias in Iraq and the surrounding countries are linked to Iran both financially and politically. One such group was the Mahdi Army. This organization emerged in direct response to the US invasion of Iraq,* and although it's no longer in operation, it spawned several other militias, including the Promised Day Brigade, which continues to receive "training, funding and direction from Iran's Revolutionary Guard Corps Quds Force."[63] The proliferation of such groups is why the former director of the CIA, David Petraeus, has argued that the biggest future threat in the Middle East isn't the Islamic State but the various militias supported by Iran.† "Longer term," he said, "Iranian-backed Shia militia could emerge as the preeminent power in the country, one that is outside the control of the government and instead answerable to Tehran."[64]

Thus, it may turn out that the most significant eschatological clash in the Middle East isn't between Christian Zionists and Muslim extremists. Instead, the primary actors could be the two branches of Islam, with the Islamic State on one side (although new Sunni terror groups could arise

---

* As David Cook further points out, the Mahdi Army "likely" held that "the purpose of the US-led invasion was to initiate an apocalyptic war—in this case, to find the Mahdi and to kill him." Thus, the Mahdi Army's "mission is either literally to defend the Mahdi from American forces or figuratively to defend the Shiite community." See "Messianism in the Shiite Crescent," Hudson Institute, April 8, 2011, http://www.hudson.org/research/7906-messianism-in-the-shiite-crescent.

† Adding yet another layer of complexity to the eschatological clash, the Lebanon-based Shia militia group Hezbollah released literature during its 2006 conflict with Israel (a war mentioned in the previous section, in connection with the Christian Zionist Janet Parshall) that prominently features the Mahdi.

just as quickly as it did) and Iran on the other (with its affiliated militias). Recall the Shia fighter quoted at the beginning of this section, who claimed that the US-led attack on Iraq was "the first sign" of the Mahdi's return. This statement comes from a Reuters article titled "Apocalyptic Prophecies Drive Both Sides to Syrian Battle for End of Time." Thus, the article's counterpoint is represented by a Sunni jihadist who states that "if you think all these mujahideen came from across the world to fight [the Syrian president] Assad, you're mistaken. They are all here as promised by the Prophet. This is the war he promised—it is the Grand Battle."[65] The clash of eschatologies is, clearly, a complex phenomenon, with people from multiple faith traditions driven by incompatible end-times beliefs coming into violent contact with each other. While this predicament may evolve in its specifics, it will almost certainly stick around as a general theme for decades, perhaps centuries, to come.

In 2015, Israel's prime minister, Benjamin Netanyahu, referred to Iran as "a messianic apocalyptic cult controlling atomic bombs." He added that "when the wide-eyed believer gets hold of the reins of power and the weapons of mass death, then the entire world should start worrying, and that is what is happening in Iran."[66] This is ironic, of course, because a significant portion of Netanyahu's supporters in the US are themselves "wide-eyed believers" with "the reins of power and the weapons of mass death." Meanwhile, sandwiched between Israel and Iran, the Islamic State is a millenarian cult with its eyes fixed on Armageddon in Syria, and the world's end in 2076. Members of this group despise not only the infidel "crusaders" from the West but also the "partisans of Ali," i.e., those who adhere to Shia Islam. The sectarian tension here is so great that many Iranians have come to believe that the Islamic State is actually an "invention" of the US for the purpose of strengthening Israel's security in the region.[67] (Some conspiracy theorists in the US have proposed the same idea, but in Iran this belief is widespread among the public.) Making matters worse, Shia militias have infiltrated Iraq and the surrounding areas, anticipating the imminent return of the Twelfth Imam—and we haven't even gotten to the Religious Zionists in Israel who believe that certain territories belong to the Jews by divine decree and that violence against their enemies is sanctioned by God.[68] Given this messy tangle of eschatological fatalism, Netanyahu may be more correct than he knows when he says that "the entire world should start worrying."

It seems that the one thing religions can agree upon is that we're living in the end times. They just can't agree on the details—and this is where, if you will, the devil is.[69]

Same Page, Different Holy Book

The power to manipulate and rearrange the world in unprecedented ways is pregnant with the question: *to what end?* Now that we have the technological know-how, what about the know-why? The New Atheists, led by Peter Boghossian, Michael Shermer, Richard Dawkins, and others, have an answer: *to maximize human well-being.* But religionists also have an answer: *to do God's will, whatever it happens to be.* The clash of eschatologies is, as we've seen, a real phenomenon in the world, and it will likely have an even greater impact as the poorly designed computers between our ears, which haven't been updated much since the Pleistocene, become increasingly surrounded by levers that, when pulled, could obliterate civilization.*

Religion is, I have argued, necessary for the existential riskologist to study, no less than advanced technology is important for the New Atheist to understand. Here's why: while the West has been steadily becoming more secular (one study from 2011, for example, concluded that religion is headed for "extinction" in nine Western countries), superstitious belief *worldwide* is on the rise.[70] According to a recent Pew Research poll, "by 2050, *more than 6 out of 10 people on Earth will be Christian or Muslim.*"[71]

This is disconcerting because, while the above discussion focused primarily on eschatological activists still mostly on the fringe, the fact is that a *huge percentage* of religious people profess that the end is just around

* It's worth noting that, moving forward, this clash will be increasingly framed by widespread food insecurity, changing coastlines, severe heat waves, megadroughts, more extreme weather, dwindling natural resources, and worsening infectious disease, as a result of global warming and biodiversity loss. Such phenomena will almost certainly exacerbate whatever social, political, and religious tensions there already are in the world. As Amanda Mayoral observes, climate change is a "conflict multiplier." In some cases, the effects of environmental catastrophe will actually serve to further reinforce the convictions of religionists that the end is nigh, since trouble, strife, suffering, wars, and natural disasters are seen as harbingers of the end. See Amanda Mayoral, "Climate Change as a Conflict Multiplier," United States Institute of Peace, Peace Brief no. 120, February 2, 2012, http://www.usip.org/publications/climate-change-conflict-multiplier.

the corner. For such individuals, *there's simply no such thing as human extinction*, and consequently the topic of existential risks is a *nonstarter.* Republican congressman John Shimkus, for example, once dismissed the dangers posed by global warming before a House Energy Subcommittee on Energy and Environment hearing in 2009 by citing God's promise to Noah after the great flood that "never again shall there be a flood to destroy the earth" (Genesis 9:11).[72] The same reasoning would presumably lead one to oppose research to locate deadly asteroids and comets in our vicinity or better understand how humanity might survive a supervolcanic eruption. God tells us we won't go extinct, so why worry? The fact is that within only a few decades, there could potentially be *billions* of people around the world who expect, with the unshakable firmness of faith, that the world will soon be transformed in accordance with a prewritten plan devised by an invisible Being and privately revealed to one or more prophets throughout history.

Consider the following data. According to a Pew poll from 2010, 41% of Americans believe Jesus will either "definitely" or "probably" return by 2050. (For some perspective, a 1993 poll found that 20% of those asked believed the Second Coming would happen circa the year 2000.) The number goes up to 58% among white evangelical Christians, the primary supporters of organizations like CUFI. Another survey reports that, among evangelical leaders around the globe, 61% believe in the Rapture. In sub-Saharan Africa, a whopping 82% of leaders accept the Rapture as a reality. Some 52% of respondents also affirmed that Jesus will return within their lifetimes.[73] With respect to US support for Israel, a Gallup poll reveals that the more religious one is, the more likely one is to support Zionist causes.[74] (And the US is *by far* the most religious country in the developed world.)

---

* As I point out in an article about being on the "right side of futurology," many atheists see theology as "a subject without an object," since God almost certainly doesn't exist. Interestingly, the exact same can be said about religious believers and Existential Risk Studies: this field is, on their view, concerned with phenomena that don't exist—that can't possibly happen. It's impossible for us to go extinct because such a scenario would violate the metaphysics and eschatology of religion. Thus, studying existential risks is as much a waste of time as studying unicorns or leprechauns. For further discussion of this point, see Phil Torres, "On the 'Right Side of Futurology': Atheism and Human Extinction," September 5, 2015, Debunking Christianity, http://debunkingchristianity.blogspot.com/2015/09/on-right-side-of-futurology-atheism-and.html.

Yet another study found that, as of 2013, 13% of the voting public thinks Barack Obama is the Antichrist.[75]

With respect to Islam, a 2012 Pew poll reports that "in nine of the 23 nations where the question was asked, half or more of Muslim adults say they believe the return of Mahdi will occur in their lifetime." This belief is highest in Afghanistan (83%) and Iraq (72%), both of which have recently witnessed great violence. Similarly, "half or more Muslims in seven nations say they expect to be alive to see [Jesus return]." This conviction is particularly widespread in Turkey (65%) and Iraq (64%).[76] These are significant results because the population of Muslims is projected to be 2.76 billion by 2050 (it's the fastest growing religion in the world). This means that even a *tiny percentage* of extremists will constitute a huge group in absolute numbers. Indeed, 1% of 2.76 billion is *27.6 million*, which is 26.2 million greater than the number of military personnel on active duty in the US. Given the dual trends of power and accessibility exhibited by certain emerging technologies, it follows that more and more people will not only *believe* that advanced technology should be used to inflict catastrophic harm on society but also be able to actually *follow through* on such a nightmare.

Notes

1.   Townshend writes: "In the 1980s terror was still the business of a handful of radical revolutionaries and some all-too-familiar nationalists. The next ten years, however, saw a remarkable shift. One of the leading surveys in the late 1990s asserted that 'the religious imperative for terrorism is the most important defining characteristic of terrorism today,' while the author of an American college textbook on terrorism put 'religious fanaticism' top of her list of terrorist motives. Official assessments reflect this too; for instance, the *Canadian Security Intelligence Service 2000 Public Report* states that 'one of the prime motivators of contemporary terrorism is Islamic religious extremism'. And while the US State Department remains unshakably regional-political in orientation, . . . its *Patterns of Global Terrorism* notes as one of the shifting trends 'a change from primarily politically motivated terrorism to terrorism that is more religiously or ideologically motivated.'" See Charles Townshend, *Terrorism: A Very Short Introduction* (Oxford University Press, 2011), pp. 97—98.

2.   See Malise Ruthven, "The Map ISIS Hates," *New York Review of Books*, June 25, 2014, http://www.nybooks.com/blogs/nyrblog/2014/jun/25/map-isis-hates/.

3.   Interestingly, "research has shown that a significant number of, but by no means all, so-called 'lone-wolf terrorists' have been suffering from a diagnosable mental illness." For more, see Kathryn Seifert, "Lone-Wolf Terrorists and Mental Illness," *Psychology Today*, January 20, 2015, https://www.psychologytoday.com/blog/stop-the-cycle/201501/lone-wolf-terrorists-and-mental-illness.

4.   See Jessica Stern and J. M. Berger, *ISIS: The State of Terror* (HarperCollins, 2015), p. 225.

5.   See Jerry Walls, "Introduction," in *Oxford Handbook of Eschatology*, ed. Jerry Walls (Oxford University Press, 2008), p. 10. The italics are mine.

6.   See Jean-Pierre Filiu, *Apocalypse in Islam* (University of California Press, 2011), p. xx. The italics are mine.

7.   See Matthew Sharpe, "On Eschatology and the 'Return to Religion,'" *Arena Journal* 39, no. 40 (2012).

8.   See Tony Campolo, "The Ideological Roots of Christian Zionism," *Tikkun*, 20, no. 1 (2005).

9.   As Donald Wagner writes, "Both Lord Arthur Balfour, author of the famous 1917 Balfour Declaration, and Prime Minister David Lloyd George, the two most powerful men in British foreign policy at the close of World War I, were raised in dispensationalist churches and were publicly committed to the Zionist agenda for 'biblical' and colonialist reasons." See "Evangelicals and Israel: Theological Roots of a Political Alliance," *Christian Century*, November 4, 1998, pp. 1020–1026.

10.   See Sam Harris, *The End of Faith* (Norton, 2005), p. 153.

11.   Thanks to Stephen Spector for pointing out this passage in his book, *Evangelicals and Israel: The Story of American Christian Zionism*, p. 27. (Personal communication.)

12.   To borrow a phrase from Bart Ehrman, *Misquoting Jesus: The Story Behind Who Changed the Bible and Why* (HarperCollins, 2005), p. 12.

13.   See Tony Campolo, "The Ideological Roots of Christian Zionism," *Tikkun*, 20, no. 1 (2005).

14.   See Michael Sells, "Armageddon in Christian, Sunni, and Shia Traditions," in *Oxford Handbook of Religion and Violence*, ed. Mark Juergensmeyer, Margo Kitts, and Michael Jerryson (Oxford University Press, 2013), p. 475.

15.   See Louis Jacobson, "Administration Is Keeping to Terms of Agreement," *Politifact*, Feburary 1, 2012, http://www.politifact.com/truth-o-meter/promises/obameter/promise/133/provide-30-billion-over-10-years-to-israel/.

16. See Daniel Schorr, "Reagan Recants: His Path from Armageddon to Detente," *Los Angeles Times*, January 3, 1988, http://articles.latimes.com/1988-01-03/opinion/op-32475_1_president-reagan.

17. Quoted from Hanna Segal, "Silence Is the Real Crime," in *Psychoanalysis and the Nuclear Threat: Clinical and Theoretical Studies*, ed. Howard Levine, Daniel Jacobs, and Lowell Rubin (Analytic Press, 1988), p. 42.

18. See Sam Harris, *The End of Faith* (Norton, 2005), pp. 153.

19. See Jerry Walls, "Introduction," in *Oxford Handbook of Eschatology*, ed. Jerry Walls (Oxford University Press, 2008), p. 14.

20. See Max Blumenthal, "Birth Pangs of a New Christian Zionism," *Nation*, August 14, 2006, http://www.thenation.com/article/birth-pangs-new-christian-zionism.

21. See "Response to Christian Zionism," National Council of Churches, http://nationalcouncilofchurches.us/common-witness/2007/christian-zionism.php.

22. See Stephen Spector, *Evangelicals and Israel: The Story of American Christian Zionism* (Oxford University Press, 2009), p. 57.

23. See Glenn Shuck, "Christian Dispensationalism," in *The Oxford Handbook of Millennialism*, ed. Catherine Wessinger (Oxford University Press, 2011), p. 525.

24. I adjusted the number of members to reflect the most recent figures. See Sulome Anderson, "The FP 50: The 50 Most Powerful Republicans on Foreign Policy," *Foreign Policy*, August 24, 2012, http://foreignpolicy.com/2012/08/24/the-fp-50-2/.

25. See Max Blumenthal, "Birth Pangs of a New Christian Zionism," *Nation*, August 14, 2006, http://www.thenation.com/article/birth-pangs-new-christian-zionism.

26. Hagee later apologized to the Jewish Anti-Defamation League.

27. See "Transcript," *Bill Moyers Journal*, October 5, 2007, http://www.pbs.org/moyers/journal/10052007/transcript1.html.

28. See John Hagee, "The Coming Holy War," *Charisma Magazine*, http://www.charismamag.com/blogs/431-j15/features/israel-the-middle-east/1818-the-coming-holy-war.

29. See Ashley Killough, "At Christmas Party, Bachmann Tells Obama to Bomb Iran," *CNN*, December 22, 2014, http://www.cnn.com/2014/12/12/politics/bachmann-bomb-iran/.

30. See "Bachmann: End Times Are Here, Thanks Obama," *Right Wing Watch*, April 13, 2015, http://www.rightwingwatch.org/content/bachmann-end-times-are-here-thanks-obama.

31. Again, see "Transcript," *Bill Moyers Journal*, October 5, 2007, http://www.pbs.org/moyers/journal/10052007/transcript1.html.

32. See Paul Boyer, "John Darby Meets Saddam Hussein: Foreign Policy and Bible Prophecy," *Chronicle of Higher Education* 49, no. 23 (2003).

33. See Mariam Karouny, "Apocalyptic Prophecies Drive Both Sides to Syrian Battle for End of Time," *Reuters*, April 1, 2014, http://www.reuters.com/article/2014/04/01/us-syria-crisis-prophecy-insight-idUSBREA3013420140401. Or, as Filiu puts it in *Apocaypse in Islam* (University of California Press, 2011), "This apocalyptic sense of urgency was aggravated by the American invasion of Iraq in 2003, which gave new life to Shi'I messianism."

34. See Spencer Ackerman, "'Apocalyptic' Isis beyond Anything We've Seen, Say US Defence Chiefs," *Guardian*, August 22, 2014, http://www.theguardian.com/world/2014/aug/21/isis-us-military-iraq-strikes-threat-apocalyptic.

35. See Scott Baldauf, "The 'Cave Man' and Al Qaeda," *Christian Science Monitor*, October 31, 2001, http://www.csmonitor.com/2001/1031/p6s1-wosc.html.

36. See Peter Chalk, *Encyclopedia of Terrorism, Volume 1* (ABC-CLIO, 2013), p. 469.

37. See Liz Sly, "Al-Qaeda Disavows Any Ties with Radical Islamist ISIS Group in Syria, Iraq," *Washington Post*, February 3, 2014, http://www.washingtonpost.com/world/middle_east/al-qaeda-disavows-any-ties-with-radical-islamist-isis-group-in-syria-iraq/2014/02/03/2c9afc3a-8cef-11e3-98ab-fe5228217bd1_story.html.

38. See Graeme Wood, "What ISIS Really Wants," *Atlantic*, March 2015, http://www.theatlantic.com/features/archive/2015/02/what-isis-really-wants/384980/.

39. See Madeline Grant, "16% of French Citizens Support ISIS, Poll Finds," *Newsweek*, August 26, 2014, http://www.newsweek.com/16-french-citizens-support-isis-poll-finds-266795, and "UK Poll Shows Up to 1.5 Million ISIS Supporters in Britain," *Clarion Project*, July 8, 2015, http://www.clarionproject.org/news/uk-poll-shows-15-million-isis-supporters-britain.

40. See "CIA Chief Says ISIS Has 'Snowballed,'" *Fox News*, March 14, 2015, http://www.foxnews.com/politics/2015/03/14/reports-success-against-isis-overinflated/.

41. See Peter Bergen, "ISIS Goes Global," *CNN*, March 8, 2015, http://www.

cnn.com/2015/03/08/opinions/bergen-isis-boko-haram/index.html.

42.   See "The Islamic State (Full Length)," *Vice News*, published August 14, 2014, https://www.youtube.com/watch?v=AUjHb4C7b94.

43.   See Graeme Wood, "What ISIS Really Wants," *Atlantic*, March 2015, http://www.theatlantic.com/features/archive/2015/02/what-isis-really-wants/384980/.

44.   The quote comes from an interview with Graeme Wood, in which he also states that the final caliph will be the Mahdi. See "Journalist Graeme Wood on the Islamic State: VICE Meets," *Vice News*, published November 7, 2014, https://www.youtube.com/watch?v=RQ4VhSMHzkk. For the article, see Graeme Wood, "What ISIS's Leader Really Wants," *New Republic*, September 1, 2014, http://www.newrepublic.com/article/119259/isis-history-islamic-states-new-caliphate-syria-and-iraq.

45.   See William McCants, "ISIS Fantasies of an Apocalyptic Showdown in Northern Syria," Brookings Institute, October 3, 2014, http://www.brookings.edu/blogs/markaz/posts/2014/10/03-isis-apocalyptic-showdown-syria-mccants.

46.   Quoted from Graeme Wood, "What ISIS Really Wants," *Atlantic*, March 2015, http://www.theatlantic.com/features/archive/2015/02/what-isis-really-wants/384980/.

47.   See Saba Abu Farha, "Many Jordanians Back ISIL Despite Pilot's Killing," *USA Today*, February 8, 2015, http://www.usatoday.com/story/news/world/2015/02/08/jordan-islamic-state-support/22988539/.

48.   See William McCants, "ISIS Fantasies of an Apocalyptic Showdown in Northern Syria," Brookings Institute, October 3, 2014, http://www.brookings.edu/blogs/markaz/posts/2014/10/03-isis-apocalyptic-showdown-syria-mccants.

49.   As David Cook notes, for Muslim apocalypticists in general, "the only way to return to the idealized state and to actualize justice in the Muslim community (umma) is through violent events that will purify the community and focus it away from internal enemies towards external ones." See David Cook, "The Mahdi's Arrival and the Messianic Future State According to Sunni and Shi'ite Apoclayptic Scnearios," Nehemia Levtzion Center for Islamic Studies, http://www.hum.huji.ac.il/upload/_FILE_1415823040.pdf, p. 5.

50.   See Alexander Sehmer, "Isis Could Obtain Nuclear Weapon from Pakistan, Warns India," *Independent*, May 31, 2015, http://www.independent.co.uk/news/world/asia/india-warns-isis-could-obtain-nuclear-weapon-from-pakistan-10287276.html. See "NTI Nuclear Materials Security Index," Nuclear Threat Initiative, January 2014, http://ntiindex.org/wp-content/uploads/2014/01/2014-NTI-Index-Report1.pdf. Incidentally, the US ranks 11.

51.   See Damien McElroy, "Islamic State Seeks to Use Bubonic Plague as a Weapon of War," *Telegraph*, August 29, 2014, http://www.telegraph.co.uk/news/worldnews/middleeast/iraq/11064133/Islamic-State-seeks-to-use-bubonic-plague-as-a-weapon-of-war.html.

52.   See Damien McElroy, "Islamic State Seeks to Use Bubonic Plague as a Weapon of War," *Telegraph*, August 29, 2014, http://www.telegraph.co.uk/news/worldnews/middleeast/iraq/11064133/Islamic-State-seeks-to-use-bubonic-plague-as-a-weapon-of-war.html.

53.   With some elements of democracy.

54.   See David Cook, "Messianism in the Shiite Crescent," Hudson Institute, April 8, 2011, http://www.hudson.org/research/7906-messianism-in-the-shiite-crescent.

55.   See Paul Hughes, "Iran President's Religious Views Arouse Interest," *Washington Post*, November 17, 2005, http://www.washingtonpost.com/wp-dyn/content/article/2005/11/18/AR2005111801625_pf.html.

56.   See Jean-Pierre Filiu, *Apocalypse in Islam* (University of California Press, 2011), p. 151.

57.   See Golnaz Esfandiari, "Iran: President Says Light Surrounded Him During UN Speech," Radio Free Europe, Radio Libery, November 29, 2005, http://www.rferl.org/content/article/1063353.html.

58.   See Rick Gladstone, "Iran's President Calls Israel 'an Insult to Humankind'," *New York Times*, August 17, 2012, http://www.nytimes.com/2012/08/18/world/middleeast/in-iran-ahmadinejad-calls-israel-insult-to-humankind.html.

59.   See Eliot C McLaughlin, "Iran's Supreme Leader: There Will Be No Such Thing as Israel in 25 years," September 10, 2015, http://www.cnn.com/2015/09/10/middleeast/iran-khamenei-israel-will-not-exist-25-years/.

60.   See Reza Kahlili, "New Iran President Thanks 'Messiah' for Victory," *WorldNetDaily*, June 23, 2013, http://www.wnd.com/2013/06/new-iran-president-thanks-messiah-for-victory/.

61.   Kalaji adds that "because of the lack of any public documents, or these adherents' incompetence in writing books or articles, many ambiguities surround their views." See Mehdi Khalaji, "Apocalyptic Visions and Iran's Security Policy," in *Deterring the Ayatollahs: Complications in Applying Cold War Strategy to Iran*, ed. Patrick Clawson and Michael Eisenstadt (Washington Institute for Near East Policy, 2007).

62.   See "Iran Defends Nuclear Secrecy," *Global Security Newswire*, April 2,

2007, http://www.nti.org/gsn/article/iran-defends-nuclear-secrecy/.

63.   See Ernesto Londoño and Karen DeYoung, "US Commanders Are Concerned about New Iraqi Restrictions on American Troops," *Washington Post*, July 18, 2009, http://www.washingtonpost.com/wp-dyn/content/article/2009/07/17/AR2009071703634.html.

64.   See Jeremy Diamond, "Petraeus: ISIS Isn't Biggest Long-term Threat to Region," *CNN*, March 21, 2015, http://www.cnn.com/2015/03/20/politics/petraeus-greatest-threat-iraq-isis-shiite-militias/.

65.   See Mariam Karouny, "Apocalyptic Prophecies Drive Both Sides to Syrian Battle for End of Time," *Reuters*, April 1, 2014, http://www.reuters.com/article/2014/04/01/us-syria-crisis-prophecy-insight-idUSBREA3013420140401. Thanks to J. M. Berger for emphasizing this clash in an email exchange.

66.   See Jeffrey Goldberg, "Netanyahu Confronts Obama, and a 'Messianic Apocalyptic Cult,'" *Atlantic*, March 3, 2015, http://www.theatlantic.com/international/archive/2015/03/netanyahu-vs-a-messianic-apocalyptic-cult/386650/.

67.   See Thomas Erdbrink, "For Many Iranians, the 'Evidence' Is Clear: ISIS Is an American Invention," *New York Times*, September 10, 2014, http://www.nytimes.com/2014/09/11/world/middleeast/isis-many-in-iran-believe-is-an-american-invention.html. See also Kay Armin Serjoie, "Why Iran Believes ISIS Is a US Creation," *Time*, February 26, 2015, http://time.com/3720081/isis-iran-us-creation/.

68.   As Nur Masalha writes, "For messianic Zionists, the conflict with 'gentiles' over Jerusalem, and even war against them, is 'for their own good,' because this will hasten messianic redemption. . . . For the messianic rabbis, who embrace the supremacist paradigms of Jews as a divinely 'Chosen People' and Israel as a sacred racial state, the indigenous Palestinians are no more than illegitimate tenants and squatters, and a threat to the process of messianic redemption; their human and civil rights are no match for the divine plan and the divine ordained commandment . . . of conquering, ethnic cleansing, possessing and settling the 'promised land.'" See Nur Masalha, *The Zionist Bible: Biblical Precedent, Colonialism and the Erasure of Memory* (Routledge, 2014), p. 200. See also Mark Juergensmeyer, *Terror in the Mind of God: The Global Rise of Religious Violence* (University of California Press, 2001), p. 150. For a quite entertaining look at some Jewish extremists, see Louis Theroux's 2011 BBC documentary *The Ultra Zionists*.

69.   For an excellent and accessible examination of the clash of eschatologies, see Paul Boyer, "The Foreordained Future: Apocalyptic Thought in the Abrahamic Religions," *Hedgehog Review* 10, no. 1 (2008).

70.   See Jason Palmer, "Religion May Become Extinct in Nine Nations, Study Says," *BBC*, March 22, 2011, http://www.bbc.com/news/science-environment-12811197.

71.   Emphasis added. See Daniel Burke, "The world's fastest-growing religion is . . . ," *CNN*, April 3, 2015, http://www.cnn.com/2015/04/02/living/pew-study-religion/.

72.   See "'The Planet Won't Be Destroyed by Global Warming Because God Promised Noah', Says Politician Bidding to Chair US Energy Committee," *Daily Mail*, November 10, 2010, http://www.dailymail.co.uk/news/article-1328366/John-Shimkus-Global-warming-wont-destroy-planet-God-promised-Noah.html.

73.   See "Evangelical Beliefs and Practices," Pew Research Center, June 22, 2011, http://www.pewforum.org/2011/06/22/global-survey-beliefs/.

74.   See Frank Newport, "Religion Plays Large Role in Americans' Support for Israelis," Gallup, http://www.gallup.com/poll/174266/religion-plays-large-role-americans-support-israelis.aspx.

75.   See Ruth Brown, "Poll: 13% Think Obama Is the anti-Christ; 29% Believe in Aliens," *USA Today*, April 3, 2013, http://www.usatoday.com/story/news/nation/2013/04/03/newser-poll-conspiracy-theories/2049073/.

76.   See "Chapter 3: Articles of Faith," Pew Research Center, August 2012, http://www.pewforum.org/2012/08/09/the-worlds-muslims-unity-and-diversity-3-articles-of-faith/.

14

Proaction and Precaution

· · · · ·

Numbers and Probabilities*

Our collective expedition into the future is involuntary. The invisible hand of time is inexorably pushing us forward, into the shadows and mists of things unknown. It's almost certainly true, according to the best available evidence, that there's no after-party once our bodies succumb to entropy. (Nor are there any moral holidays for atheists, which makes ethical behavior that much more important.) And without creatures like us—the only authors of value, creators of purpose—the universe would be utterly meaningless. So the stakes are high.

Having now explored the secular and religious branches of eschatology, as well as how they're interacting to push us closer to the precipice of disaster, we're finally in a position to ask: is there reason to be optimistic about the future? The world-renowned cognitive scientist, Steven Pinker, argues in his book *The Better Angels of Our Nature: Why Violence Has Declined* that, our untrained intuitions aside, violence has actually declined throughout human history. That is to say, you're less likely to get murdered, raped, and even spanked (as a child) today than in the past. The atheist historian Michael Shermer elaborates this theme in *The Moral Arc: How Science and Reason Lead Humanity toward Truth, Justice, and Freedom*, which argues that scientific reason has led to appreciable moral progress with respect to, for example, the rights of women, gays, and even animals. The result is a rather sanguine picture of the future: things are, on the whole, getting better. But is this conclusion warranted given the phenomena discussed

* The title of this chapter is a reference to two principles, namely the proactionary principle and the precautionary principle. The first was proposed in response to the second by the transhumanist philosopher Max More.

throughout this book? When advanced technologies, some of which are becoming increasingly accessible to single individuals, are injected into the picture, what results? How optimistic should we really be about our technological future? And are there things we can do to actively inflate the probability of, as Bostrom puts it, an "okay outcome" for intelligent life on Earth?[1]

To answer these questions rigorously, we need to make explicit several recent big-picture risk trends. The first concerns the number of *possible disasters*, which has undergone a growth spurt in recent decades. Consider the fact that less than a century ago, there were perhaps three or four plausible ways our species could have kicked the bucket. For example, we could have died out from a supervolcanic eruption, an asteroid or comet impact, or (maybe) a global pandemic. Other possible existential risks from nature, not discussed earlier, are supernovae, black hole explosions or mergers, galactic center outbursts, and gamma-ray bursts. All of these are, as far as we know, still with us today. They constitute what we might call our *cosmic risk background*.

Since at least the mid-twentieth century, though, the number of risks has grown. Today, there are *far more* doomsday scenarios than ever before in history, most of these being anthropogenic in origin. It's hard to specify an exact number, but if the contemporary riskologist were pushed to give one, she might say there are *at least 20* scenarios either presently before us or ominously lingering in the foreseeable future. Only a handful of what I take to be the most pressing risks were examined in this book (with the exception of those discussed in chapter 11, which aren't urgent). Additional possibilities include a totalitarian takeover; a gradual loss of human fertility; a dysgenic scenario, resulting in significant losses of human intelligence; a physics experiment that causes a negatively charged strangelet reaction or a catastrophic vacuum decay, which could literally destroy the entire universe; an extraterrestrial invasion, which may be more likely today because of leaked radiation from Earth and Active SETI (see below); large-scale cyberterrorism; and even a successful propaganda campaign by the Voluntary Human Extinction Movement, which is unlikely, but possible.

It follows that in less than a century, humanity has brought about a sudden and rapid proliferation of *existential risk scenarios*. There are many more ways for our species to skip from this side of the grave to the other than in any prior period, largely because of our phenomenal success as inventors, scientists, and engineers. This is an extremely important point, one that should make us quite cautious when talking about how "things are getting better."

But more ways to perish doesn't necessarily mean that the probability of an existential catastrophe is higher. There could, for example, be only a single annihilation scenario that's virtually certain to occur, or a multiplicity of scenarios that are each highly improbable. Nonetheless, the very best estimates by the most qualified riskologists today suggest that the probability of doom has indeed risen over the past few centuries. The result is a kind of uniformity of distress among the experts about the future of humanity. Philosopher Bertrand Russell and Albert Einstein described a similar situation back in 1955, in their coauthored "manifesto" urging nations "to find peaceful means for the settlement of all matters of dispute." Writing specifically about nuclear annihilation, they said: "Many warnings have been uttered by eminent men of science and by authorities in military strategy. None of them will say that the worst results are certain. What they do say is that these results are possible, and no one can be sure that they will not be realized. We have not yet found that the views of experts on this question depend in any degree upon their politics or prejudices. They depend only, so far as our researches have revealed, upon the extent of the particular expert's knowledge. We have found that *the men who know most are the most gloomy.*"[2] This statement is, I believe, more or less applicable to the field of existential risks as a whole (which of course includes the topic of nuclear annihilation).[3] Those who know the most tend to be the most anxious about our survival.

Consider, for example, the influential 2006 Stern Review on the Economics of Climate Change, which describes global warming as "the greatest and widest-ranging market failure ever seen." Led by the economist Sir Nicholas Stern, this review concludes that there's a 9.5% chance of humanity going extinct *this century.*[4] Similarly, a survey of big-picture risks taken during a Future of Humanity Institute–hosted conference on global catastrophic risks (discussed more below) places the likelihood of total annihilation this century at 19%.[5] Bostrom, one of the pioneers of Existential Risk Studies, writes that assigning a probability of less than 25% "would be misguided," adding that "the best estimate may be considerably higher."[6] And the Astronomer Royal and cofounder of the Centre for the Study of Existential Risk (CSER), Sir Martin Rees, writes in his sobering book *Our Final Hour* that our species has a mere 50% chance of making it through the twenty-first century[7]—a pitiful coin toss!

Many other experts have proposed estimates in the same general probability neighborhood.* My own considered opinion aligns more or less with Rees' figure, although if anything I'm inclined to see his estimate as *slightly too sanguine*, in part because of considerations discussed in chapter 10 about cognitive closure. To be sure, the annihilation of every last person on the planet would be an extraordinary event: think about the uncontacted tribes in the Brazilian Amazon, the small towns in northern Siberia, and the roughly 60 people living in Las Estrellas, Antarctica. But just as surely, it's more and more the case that the mechanisms of death needed to do precisely this are dropping into the hands of more and more people.

The takeaway idea here is that humanity has also brought about a sudden and rapid increase in the *likelihood of a worst-case scenario being realized*, from a negligible probability for most of our species' history to a probability perhaps as high as 50% in the next 100 years. As the saying, sometimes attributed to Yogi Berra, goes: "The future ain't what it used to be!"

Two Hypotheses about the Existential Threat

What should we make of such observations, of our evolving existential predicament? What should we infer from these unsettling trends? We're increasingly surrounded by risks of the global-transgenerational (red-dot or black-dot, in figure B) variety, and the most knowledgeable experts think our chances of escaping catastrophe have significantly *dropped* since World War II. Are we merely passing through a phase of elevated eschatological threats? Or could we be approaching an ineluctable Great Filter that's guaranteed to snap off the only remaining twig on the once-bushy branch of our genus, *Homo*? Could the collision between the archaic beliefs held by so many people and the neoteric technologies invented by our best and brightest result in an unavoidable disaster?

First, with respect to the growing number of existential risk scenarios, we can say with a high degree of confidence that (i) *if* the development of increasingly powerful technologies continues into the future, and (ii) *if* such technologies are dually usable, then brand new anthropogenic

* The philosopher John Leslie, an early pioneer of Existential Risk Studies, assigns a 30% probability of humanity perishing in the next five centuries in his book **The End of the World: The Science and Ethics of Human Extinction** (Routledge, 1996).

scenarios will continue to be created. At present, there are a few reasons that condition (i) might not occur. The most obvious is that an existential catastrophe happens, resulting in irreversible technological stagnation or the permanent end of innovation. Other possibilities include future people losing interest in technology and a totalitarian state with techno-conservative views gaining global power.

In the absence of such defeaters, though, it seems extremely likely that technological development will continue along present trajectories. The fact is that technology has become something of an *autonomous phenomenon*.[8] Its development is dependent upon the actions of individuals, of course, but no single individual, organization, or government actually controls it. We're like starlings in a flock (called a "murmuration") that's darting this way and that: the flock's movement ultimately depends upon the movement of individual starlings, but no single starling is in charge of its direction. There's a sense in which the flock moves *independently* of the individuals. The exact same can be said of technology. This is why calls to "relinquish" particular areas of research, as the cofounder of Sun Microsystems, Bill Joy, famously did in a much-discussed *Wired* article from 2000, are misguided and ultimately otiose. The realization of certain technologies is more or less *inevitable*, as long as the defeaters above don't occur.[9] (We might call this "*ceteris paribus* inevitability.")

Looking at condition (ii), I see no reason for expecting future technologies to be any less dually usable than those around us today. It appears to be an intrinsic feature of technological design that there will always be *some* wiggle room with respect to an artifact's application: no matter how well-designed a laptop is, for example, it can always be dropped from a ten-story window to kill someone below. Similarly, a handgun can be used as a paperweight or a doorstop, even though it's specifically designed to injure and kill. It's not even clear what a mono-use technology would look like, especially, perhaps, in domains like synthetic biology, nanotechnology, and artificial intelligence. The artifacts of the future will, therefore, almost certainly be no less dually usable than the centrifuges that enrich uranium and the petri dishes in which bacteria cultures grow. Putting these two premises together, then, it follows that the future will be increasingly populated by new kinds of risk scenarios. We started out with three or four a few centuries ago, are now confronted by about twenty, and will very likely encounter even more in the future.

But what about the probability? What about the *threat* of an existential catastrophe? As mentioned, more risk scenarios doesn't *necessarily* mean a higher chance of death, although there may be reasons for thinking that, in our particular situation, it does. Here we can distinguish between two hypotheses about the future threat posed by new existential risk scenarios.[10] We'll call the first the "bottleneck hypothesis." It claims that the recent increase in apocalyptic dangers is a mere blip, a temporary and passing phase of heightened hazards. Once we get beyond this period, everything will be fine: technological development will continue (thus making life better and better) while the probability of disaster will subside, perhaps even dipping below pre-1945 levels.*

The bottleneck hypothesis is perhaps the most commonly held view. It finds expression in phrases like, "Humanity is at a crossroads" and "This is the most important century in human history." Some of the most prominent riskologists, such as Rees, Bostrom, and Kurzweil, have championed this hypothesis in their work, if only implicitly. For example, Rees writes that "our choices and actions could ensure the perpetual future of life (not just on Earth, but perhaps far beyond it, too). Or in contrast, through malign intent, or through misadventure, twenty-first century technology could jeopardise life's potential, foreclosing its human and posthuman future." Rees continues by declaring that "what happens here on Earth, in this century, could conceivably make the difference between a near eternity filled with ever more complex and subtle forms of life and one filled with nothing but base matter."[11]

Bostrom echoes this sentiment, saying: "One might argue . . . that the current century, or the next few centuries, will be a critical phase for humanity, such that if we make it through this period then the life expectancy of human civilization could become extremely high."[12] He later repeats this idea in a 2013 paper, which argues that "there are many reasons to suppose that the total such risk confronting humanity *over the next few centuries* is significant" (italics added).[13] The bottleneck hypothesis also informs Bostrom's quasi-poetic "Letter from Utopia," in which he describes a paradisiacal world of posthumans marked by a profusion of indescribable pleasure.[14]

* For example, we could use advanced technologies to destroy asteroids/comets heading for Earth, to lessen the impact of supervolcanic eruptions, and to develop vaccines for every infectious disease nature throws at us.

As for Kurzweil, the bottleneck hypothesis is more or less built into his singularitarian eschatology, which we haven't had occasion to discuss in this book. Basically, Kurzweil claims that after the Singularity (when "technological change [becomes] so rapid and profound it represents a rupture in the fabric of human history"), the universe will "wake up" and "the 'dumb' matter and mechanisms of the universe will be transformed into exquisitely sublime forms of intelligence, which will constitute the sixth [and final] epoch in the evolution of patterns of information."[15] He adds that this highly desirable state "is the ultimate destiny of the Singularity and of the universe."[16]

In a phrase, the bottleneck hypothesis asserts that *if only we can make it through the current squeeze*, things will settle down and our future will become more secure. We may even enter into a new, techno-utopian state of endless bliss.

The alternative hypothesis rejects this as confusing growth pains with the aches of terminal cancer. We can call this the "parallel growth hypothesis." On this account, the historical trend concerning the probability of catastrophe is no less projectable into the future than any other trend, such as the development of computer hardware described by Moore's Law. In other words, the existential threat will *track* the developmental trajectory of advanced dual-use technologies. If this growth is linear, then the existential threat will grow linearly. If this growth is exponential, then we might expect something like an *existential risk singularity*, or a future period during which the pace at which new risk scenarios are introduced becomes so rapid that the probability of total annihilation quickly approaches 1.* Either way, the central idea is that future technologies will introduce so many new ways for our species to destroy itself that an escape from doom will become practically impossible. To borrow a line from Franz Kafka, there may be "plenty of hope, an infinite amount of hope—but not for us."

The parallel growth hypothesis is obviously less desirable than the bottleneck hypothesis. Yet I take it to be the default hypothesis of the future, unless we take a number of time-sensitive, critical steps toward changing our moral and cognitive predicament. This leads to possibly the most important part of this book, at least ethically speaking: some practical

* This is meant to mirror Ray Kurzweil's definition of the "technological Singularity" as "a future period during which the pace of technological change will be so rapid, its impact so deep, that human life will be irreversibly transformed."

strategies for maximizing the likelihood of a positive outcome for our species.[17] There are, indeed, actions we can take to *make* the bottleneck hypothesis true, even if it's not the most probable hypothesis initially. Unfortunately, the subfield of "existential risk mitigation strategies" is woefully underdeveloped, and consequently there aren't many shoulders upon which to stand. The following section, therefore, offers a mere baby step in the direction of securing a better future for our children; more work of greater sophistication needs to be done. While a few of the strategies proposed have been discussed by scholars before, others haven't—or at least not in detail. Many of the same strategies for reducing the threat of anthropogenic risks can, we should note, also help mitigate the nonanthropogenic risks posed by nature, with respect to which the error/terror distinction doesn't apply.

Strategies for Survival

1. Redesign the Human Form.
If the history and comparative study of religion teach us one thing, it's that the primate species known as *Homo sapiens* is extremely susceptible to certain kinds of delusions. The brief explorations of chapters 1 and 13 are merely the tip *of* the tip of the iceberg. Hallucinations about invisible forces, miraculous events, and supernatural beings are ubiquitous across cultural space and time, to the point that one who studies such phenomena might conclude that there's simply no hope for our species in a world full of new annihilation mechanisms. If it weren't for the fact that apocalyptic artifacts were unavailable to past humans, we'd almost certainly be goners already. Looking into the future, it's clear that, with the advent of biotechnology, synthetic biology, molecular manufacturing, and even artificial intelligence, just *one bad apple* is all it will take to ruin the party of life for *everyone*. What are the chances that we could survive such a situation? Our species simply can't be trusted to handle advanced dual-use technologies. We aren't responsible enough to have objects of this sort within reach; we're like children playing with matches, except that instead of matches we have flamethrowers.

The philosophical position known as transhumanism has an idea for how to fix this situation. Perhaps it's the case, the transhumanist might suggest, that *to survive, we must go extinct*. This may sound paradoxical, but it's not, or at least not necessarily. Recall from chapter 7 that there are two kinds of extinction: on the one hand, a species could go extinct by dying out, as happened to the dodo and (most of) the dinosaurs. On the other, a species could go extinct by evolving into another life-form. In the latter case, the species' lineage persists, but the new population of organisms becomes sufficiently different from the old to be considered a novel "kind." This is precisely the sort of human extinction that transhumanists hope for.[18] On their view, we have a moral obligation to use technology to transform ourselves inside and out, to take control of our evolutionary trajectory by modifying our phenotypic properties in desirable ways. In particular, advanced technologies could be used to create a future generation of cognitively enhanced cyborgs with the intellectual maturity, as it were, necessary to live in a world of greater existential risk *possibilities* without also increasing the *probability* of a catastrophe. By analogy, many children outgrow their obsession with fire and matches as they become adults; perhaps something similar needs to happen on the species level. We can call this option the "hope of transhumanity."

Note that such cyborgs wouldn't need to be superintelligent.* There could be ways of tweaking our cognitive architectures to make us, for example, less susceptible to apocalyptic delusions and more responsive to evidence in constructing our normative worldviews of reality (see appendix 4). This is, of course, far easier suggested than done, but the possibility remains that technologies within the next decade will yield breakthroughs in our understanding of the nervous system that enable us to intervene in expedient ways.† This would probably need to happen soon, though, since the threat of annihilation is growing rapidly. In the end, it may be that what saves us from a "bad" extinction event is a "good" extinction event—an evolutionary transition that fulfills the transhumanist goal of redesigning the human form.[19]

2. Create Superintelligence.
Another strategy, clearly related to the one above, involves the creation of one or more superintelligent beings, most likely artificial in constitution. As Bostrom writes, "One might believe that superintelligence will be developed within a few centuries, and that, while the creation of superintelligence

---

* Although recall from chapter 5 that a quantitative superintelligence could potentially reverse the ignorance explosion that's been ongoing since the Scientific Revolution.

† Scientists have, in fact, already succeeded in making "compassion pills" that can increase the empathy people feel for each other.

will pose grave risks, once that creation and its immediate aftermath have been survived, the new civilization would have vastly improved survival prospects since it would be guided by superintelligent foresight and planning."[20] In other words, as previously discussed, if a superintelligence isn't the worst thing to happen to us, it could very well be the best, ushering a new world of great value, prosperity, and existential security.

Kurzweil, famous for his singularitarian optimism, says something similar. He writes that "the window of malicious opportunity for bioengineered viruses, existential or otherwise, will close in the 2020s when we have fully effective antiviral technologies based on nanobots. However, because nanotechnology will be thousands of times stronger, faster, and more intelligent than biological entities, self-replicating nanobots will present a greater risk and yet another existential risk. The window for malevolent nanobots will ultimately be closed by strong artificial intelligence." He adds that "not surprisingly, 'unfriendly' AI will itself present an even more compelling existential risk," yet he remains thoroughly committed to his vision of superintelligent machines leading to the "ultimate destiny" of the universe.[21]

In sum, our predicament might improve in the future if we succeed in creating a superintelligence willing to help us manage the novel risk scenarios posed by advanced technologies. We can call this the "hope of posthumanity." Given that both the dangers and potential benefits associated with this strategy are *enormously huge*, I would argue that one of the best uses of current resources would be for funding research on the amity-enmity and indifference problems, which together fall under the umbrella of what Yudkowsky calls *Friendly AI*. The organizations working on Friendly AI right now are, most notably, the Machine Intelligence Research Institute (MIRI) and the Future of Humanity Institute. We'll return to this in the following section.

3. Climb into the Heavens.
This is probably the most promising way of preventing an existential catastrophe, all things considered. The idea is simple: the *wider* we spread out in the *world*, the less chance there is that a single event will have *worldwide* consequences. A collapse of the global ecosystem on Earth, for example, wouldn't affect people living on Mars, just as a grey goo disaster on Mars wouldn't affect those living on this planet. Indeed, one could imagine whole planets being wiped out without the human population as a

whole being destroyed, just as individual empires can rise and fall without global civilization collapsing. The greater the extent to which we colonize the galaxy, our Local Group, and the visible universe, the more catastrophes like nuclear war and engineered pandemics (as well as natural risks like supereruptions and asteroid impacts) could occur without threatening total annihilation.

Although space colonization may sound far-fetched to some ears, many respectable thinkers see it as an integral aspect of long-term human survival. For example, then NASA administrator Michael Griffin noted in 2006 that "human expansion into the solar system is, in the end, fundamentally about the survival of the species."[22] Stephen Hawking expressed the same idea when he said that he doesn't "think the human race will survive the next thousand years, unless we spread into space."[23] And the head of SpaceX, Elon Musk, has asserted that "there is a strong humanitarian argument for making life multi-planetary . . . in order to safeguard the existence of humanity in the event that something catastrophic were to happen."[24]

While space colonization would insulate us against a large number of current existential risks, it's worth noting that there are some it would fail to eliminate. A physics disaster, for example, could have consequences that are *cosmic* in scope. One possibility here, alluded to earlier, is that the universe isn't in its most stable state. A high-powered particle accelerator could potentially tip the balance, causing the universe to undergo a "catastrophic vacuum decay, with a bubble of the true vacuum expanding at the speed of light," destroying everything in its path.[25] Another possibility is that a species of bellicose extraterrestrials with a huge military might conquer our colonial descendants one planet at a time, until every last person has been massacred.

Given the (i) minimal risks, (ii) minimal costs,* and (iii) potential gains of colonizing space, this ought to be among the top goals of those concerned about existential risks. If we want a future for our children, humanity's motto ought to be, "We must colonize to thrive."[26]

4. Stagger the Development of Technology.[27]

It's better to have a vaccine in hand before a pathogen spreads, just as it's better to have an antimissile system in place before a missile is launched your direction. Insofar as it's possible to anticipate the creation of certain

---

* For example, it doesn't require that either of the first two recommendations are fulfilled.

technologies—contagions with synthetic genomes, self-replicating nanobots, mean-spirited superintelligences—we could in theory build the necessary defenses to protect ourselves first.

For example, rushing the construction of a global nanoimmunity shield to arrive before the technology needed to create self-replicating nanobots could mitigate the risk of ecophagy. And given that nanotechnology and superintelligence are both likely to arrive at some point, it would almost certainly be better to have superintelligence first (as the second recommendation above implies).[28] The reason is that there doesn't appear to be any way for advanced nanotechnology to mitigate the risks posed by superintelligence, but a superintelligence, if friendly, could very likely help us solve the dangers posed by nanoscale machines. The same could be said about superintelligence and the emerging fields of biotechnology and synthetic biology: the potential of one to mitigate the risks of the other is wholly asymmetrical. Thus, in the end, there may be a way to *slalom* around the threats before us, since we're whooshing downhill anyways.

5. Improve the Educational System.
A less speculative way of mitigating some of the risks before us is to improve the educational system. Education is a kind of noninvasive form of cognitive enhancement, one that focuses on improving the mental software running on our brains rather than the underlying neural hardware. With respect to terrorism in particular—one of the major sources of apocalyptic anxiety moving forward—one might think that if people were better educated, they'd be less likely to engage in terroristic activities. After all, we know that education level in the US is negatively correlated with crime: people who are educated are statistically less likely to break the law. As one study found, "A one-year increase in average years of schooling reduces murder and assault by almost 30%, motor vehicle theft by 20%, arson by 13% and burglary and larceny by about 6%."[29] And what is terrorism but a particularly menacing form of criminality?

Yet the available data show that a negative correlation between education level and terrorism doesn't hold. This is a surprising fact: terrorism is primarily a hobby of the educated middle or upper classes. For example, out of 75 terrorists who perpetrated recent attacks against Westerners, the majority had gone to college. As one article notes, "In the four attacks for which the most complete information about the perpetrators' educational levels is available—the World Trade Center bombing in 1993, the attacks on the American embassies in Kenya and Tanzania in 1998, the 9/11 attacks, and the Bali bombings in 2002—53 percent of the terrorists had either attended college or had received a college degree."[30] Indeed, the current leader of al-Qaeda, Ayman al-Zawahiri, is an eye surgeon, and the current caliph of the Islamic State, Abu Bakr al-Baghdadi, is thought to have earned a PhD from the Islamic University of Baghdad.[31]

So, education isn't an effective way to mitigate terrorism. Or might there be something wrong with the way we're looking at these data? Perhaps the correlations considered are too general—perhaps what matters isn't education *per se*, but the *sort* of education one receives. In many of the cases mentioned above, the terrorists went to school for "technical subjects like engineering." (Or, in the case of al-Baghdadi, his degree was in Islamic studies.[32]) While professions like engineering and medicine require factual knowledge, practical skills, and problem-solving acumen, they aren't centered on *critical thinking*, or the mental exercise of reflecting upon assumptions, interrogating arguments, questioning reasons, and analyzing concepts like truth and evidence. (The same can be said about theology.)

As appendix 4 explains, these activities are the core themes of epistemology. Critical thinking is, in fact, nothing more than applied epistemology: it involves putting the principles of epistemology into practice in the real world. This is an integral part of Peter Boghossian's powerful discussion of "Street Epistemology" in his book *A Manual for Creating Atheists*.[33] Boghossian offers a number of "interventions and strategies" for talking religious individuals out of their religious beliefs and replacing them with a worldview that's based on the best available, checkable evidence.*

The point is that an education in epistemology could make a real difference in the world. Indeed, scientific studies have demonstrated that critical thinking is antithetical to religious dogma. One experiment published in *Science* found that "analytic thinking promotes religious disbelief."[34] This is not merely a correlative claim, but a *causal* one: thinking critically *makes* one more likely to dismiss religious ideas as nonsense. Although secularists may find this conclusion unsurprising, it's notable that a controlled study was able to establish a causal link between good cognitive hygiene, so to speak, and an unwillingness to accept faith-based

---

* In my view, Boghossian's book is one of the best in the New Atheist literature. I highly recommend reading it, perhaps in combination with appendix 4.

propositions privately revealed to prophets claiming to have special access to the supernatural. Perhaps this connection is why the Republican Party of Texas literally wrote into their official 2012 platform that they "oppose the teaching of . . . *critical thinking skills* and similar programs" (italics added).[35]

The connection between critical thought and religiosity has nontrivial implications for our species' survival. Since most terrorism today is religious terrorism, it follows that eliminating religion (through voluntary apostasy, i.e., by teaching people to think critically) would significantly reduce the threat of terrorism. This is why I *strongly advocate* that we teach an Applied Epistemology 101 course in high schools, and/or make it a requirement for every college around the planet. Appendix 4 could serve as a template for such a course. Just imagine if citizens of the world based their views on the best available checkable data! Just imagine how much better the human condition would be if people's opinions tracked the evidence rather than being glued to immutable dogmas! These aren't platitudes. Applied epistemology could be an effective prophylactic against the destructive side of applied eschatology.

6. Overcome Anti-intellectualism.

The ideological posture of anti-intellectualism combines the closed-minded attitude of "Don't tell me the facts, I don't want to know them" with the stubborn stance of "Even if you do tell me the facts, they won't change my mind." Anti-intellectualism values faith over evidence, and it considers the opinions of nonexperts to have the same clout as those "in the know." It discourages the most vital and important of human capacities: *curiosity*, which lies at the very heart of critical thinking. Because of this, anti-intellectualism becomes a self-reinforcing cognitive dead end: it exhorts one *not* to question, *not* to be curious about other points of view, and *not* to think carefully about claims assumed to be true, all of which one must engage in to see the folly of anti-intellectualism.

This poor posture of the mind is ubiquitous in the contemporary US. While it makes an appearance on both sides of the political spectrum, it is, without a doubt, most highly concentrated on the right, among religious conservatives. Consider the most watched news network in the US, Fox News, whose motto, "Fair and Balanced," offers an instance *par excellence* of what George Orwell called "doublespeak." At least seven studies to date have shown that Fox News viewers are more misinformed than any other demographic about a range of important issues, from health care and the Iraq War to the 2010 midterm elections and global warming.[36] One study, for example, found a consistent relation between how often one watches Fox News and how misinformed one is, while another revealed that people who watch no news at all are actually *less* misinformed than fans of Fox News.[37] In a 2015 article for the *Foreign Review*, the conservative historian Bruce Bartlett describes the "profound political implications" of Fox News as "self-brainwashing."[38] The point is that Fox News has become a giant engine of anti-intellectualism: purveying misinformation, encouraging faith, and giving nonexperts a megaphone through which to shout their factually unsound opinions.

The ubiquity of anti-intellectualism among half the American population is nothing to sneeze at: it constitutes a genuinely dangerous situation. For example, if one examines the causes and proposed solutions to the biggest problems facing humanity, one consistently finds the conservative Right *denying the causes* and *obstructing the solutions*. Scientists have, for example, repeatedly warned that overconsumption, socioeconomic inequality, greenhouse gas emissions, and the consequences of "market fundamentalism" (to borrow a term from Naomi Oreskes and Erik Conway) could push civilization up to, and beyond, the brink of collapse.[39] Yet Republicans stubbornly insist on policies that encourage overconsumption, exacerbate rich-poor disparities, and hinder the adoption of sustainable technologies. Given the stakes—not to mention the closing windows of time in which meaningful action can be taken to avert certain catastrophes—this is a bad place to find ourselves.*

7. Make Sure That Women Are Involved.

Sexism remains a significant problem today, even in the most advanced sectors of the West. This is not a trivial point: just over half of the human population has been systematically prevented from engaging in science, philosophy, medicine, and so on. At present, women still earn less than

---

* We should note here that being an **intellectualist** isn't the same as being an **intellectual**. The former has to do with not cognitive capacity but cognitive propensity. That is to say, intellectualism is about curiosity, a longing for truth (even when it's incompatible with deeply held, prior beliefs), and fallibilism over dogmatism rather than raw mental abilities. While I hesitate to call myself an intellectual, I would rush to declare that I am, through and through, an intellectualist.

their male counterparts for the same jobs in many developed countries. Even in one of the most progressive slivers of US society—academia— sexism continues to disadvantage 51% of the human population. For example, a recent study by researchers at Yale University asked professors of biology, chemistry, and physics at six major universities to evaluate a recent graduate looking for a managerial position in a lab. The professors were all sent the exact same description, but half the applications had the name "John" at the top, while the other half had "Jennifer." Amazingly, both male and female professors—which goes to show how deeply ingrained sexism is in our culture—consistently favored John over Jennifer. Not only was John ranked higher (on a scale of seven) but he was offered a starting salary of roughly $4,000 more than his female doppelgänger.[40]

The imperative to include women goes beyond moral considerations of gender equality. Avoiding a doomsday scenario is necessarily a group effort, and it appears that groups themselves can exhibit something analogous to the general intelligence had by individuals. What makes a group smart? What determines its IQ? Bostrom notes that "a system's collective intelligence could be enhanced by expanding the . . . quality of its constituent intellects."[41] The idea that a group's ability is somehow a product of the abilities of its members is intuitive—surely a collection of geniuses will outperform a group of average intellects—but it turns out to be empirically false. A 2010 study published in *Science* found that there exists an insignificant link between the intelligence of group members and the intelligence of the group itself. The only aspect of members that correlates strongly with collective smarts is "social sensitivity," or the ability of individuals to "perceive each other's emotions."[42]

The reason is that groups composed of socially sensitive individuals tended to communicate more, and therefore collaborate better. Groups in which a single individual, however bright, monopolized the conversation tended to have lower IQs. The connection here to women is that, since they're statistically more socially sensitive than men, groups with more women tended to do better on a variety of cognitive, problem-solving tasks. In other words, the number of women in a group can be used to strongly predict its IQ.[43]

A high-IQ collection of individuals is more likely to solve big-picture problems—including those posed by existential risks—than a low-IQ group, of course. Since groups with lots of women tend to have higher

IQs than those with few women, it follows that women ought to play a significant role in the effort to avoid existential risks.

8. Burrow into the Earth.
This may sound like something a survivalist on National Geographic's *Doomsday Preppers* would propose, but there are sophisticated versions of it that are worth taking seriously. Indeed, a number of respectable intellectuals have suggested building large bunkers in which to seek refuge if a catastrophe were to occur. The economist Robin Hanson, for example, has proposed that 100 or more people remain in an underground residence to repopulate the planet if a disaster breaks out of the cage.[44] On a related note, the Norwegian government recently spent $9 million building a mountain bunker of sorts—called the Svalbard Global Seed Vault—to serve as "a backup for the world's 1,750 seed banks," in case a disaster were to occur.[45] Storing the seeds is paid for partly by the Global Crop Diversity Trust, funded by institutions like the Bill & Melinda Gates Foundation. As of this writing, the vault contains 863,969 samples.

9. Support Organizations Working to Understand Big-Picture Hazards.
This is perhaps the most obvious option for activism. Such organizations include the Center for the Study of Existential Risk, the Future of Humanity Institute, NASA, the Future of Life Institute, the Machine Intelligence Research Institute, the Institute for Ethics and Emerging Technologies, the Foresight Institute, the Center for Responsible Nanotechnology, the Global Catastrophic Risk Institute, the Lifeboat Foundation, and my own X-Risks Institute (www.risksandreligion.org). These institutes are home to some of the brightest minds working to secure a future for our children. But many are severely underfunded. As Bostrom observes, far more academic papers are published about dung beetles and *Star Trek* than existential risks. This is not to disparage such research (which may indeed have some value), but simply to point out that our *priorities* should be inverted, since the continuation of research on insect life and cultural artifacts is *dependent upon* our avoiding existential risks. Even a subexistential catastrophe could severely impede research in such fields.

10. Reduce Your Environmental Impact.
While this recommendation applies to everyone around the globe, of course, it's here directed primarily at wealthy persons and nations, since

those with money tend to have a far greater impact on the environment than the poor.

Consider the fact that, as of 2014, the richest 85 people in the world held the same amount of wealth as the bottom 50%.[46] This vast amount of accumulated wealth is used by the elite to consume huge quantities of limited resources. For example, "in 2005, the wealthiest 20% of the world accounted for 76.6% of total private consumption," whereas the poorest 20% accounted for a mere 1.5%. Similarly, "the poorest 10% accounted for just 0.5% [while] the wealthiest 10% accounted for 59% of *all* the consumption" (italics added).[47] Another study found that "more affluent households have a greenhouse gas effect from travel and transport that is 250 times greater than individuals in the lowest economic classes."[48] And a 2007 study from Australia found that the rich pollute more than twice as much as the poor: 58 tons of greenhouse gases per year versus 22.[49] Such correlations between wealth and pollution hold just as well on the regional level, as figure D shows. The moral catastrophe of this predicament is made even worse by the fact that poor nations will almost certainly suffer the effects of global warming and biodiversity loss far more than rich countries. This is the central issue of *climate justice*, as discussed in chapter 8.

The United States, a major world polluter, is currently one of only four countries in the world that's refused to abide by the Kyoto Protocol, an international treaty designed to keep atmospheric greenhouse gases at safe

Figure D. The Ecological Footprint by Region[50]

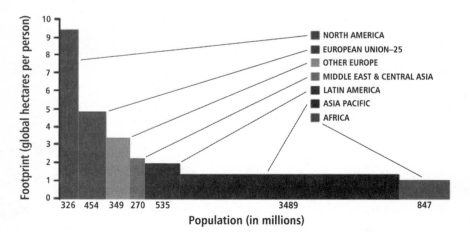

levels. The treaty has been ratified by 192 countries but not the US. It was in fact reported that George W. Bush ended a private meeting at his last G8 summit with the following bit of flippant jocularity: "Goodbye from the world's biggest polluter!"[51] And although Obama has taken a few small steps toward mitigating global warming, he hasn't done nearly enough to avert a catastrophe.

As for the individuals who pollute most—the wealthy—studies show that they are more likely to cheat, steal, lie, and endorse unethical behavior than poor people.[52] So it's unlikely that the moneyed class will reduce their impact on the environment unless doing so happens to align with their personal self-interest. The same can be said about corporations, whose policies are, in capitalist economies, fixed by the short-term motive of maximizing profit rather than the long-term desideratum of ensuring a sustainable planet for future generations. In sum, we need to reduce our environmental impact, but the actors most responsible for the plight of the biosphere today are the least likely to change their behavior.*

11. Get Overpopulation under Control.
The fact is that many of the largest-scale problems facing humanity—including water scarcity, deforestation, soil contamination, global warming, and biodiversity loss—are causally linked to the exponential growth of the global population. It's almost impossible to wrap one's head around the fact that there were only about 200 million people *in total* when Jesus was born. By 1975 this number had ballooned to roughly 4 billion, and current estimates put the number around 7 billion. We are like a rapidly growing family of twelve trying to live on a 5 x 5–foot plot of land. As the *Living Planet Report* notes, "The dual effect of a growing human population and high per capita Footprint [is multiplying] the pressure we place on our ecological resources" to the point of disaster. At current rates of consumption, we would need not 1 but 1.5 Earths to survive.[53]

---

* As a paper on the sixth mass extinction event observes, "Avoiding a true sixth mass extinction will require rapid, greatly intensified efforts to conserve already threatened species and to alleviate pressures on their populations—notably habitat loss, overexploitation for economic gain, and climate change. All of these are related to human population size and growth, which increases consumption (especially among the rich), and economic inequity" (italics added). See Gerardo Ceballos et al. "Accelerated Modern Human-Induced Species Losses: Entering the Sixth Mass Extinction," **Science Advanced** 1, no. 5 (June 2015).

It's of course possible that future technology will save us from a population collapse, just as the Green Revolution averted a Malthusian disaster in the twentieth century. But one should note that the Green Revolution introduced many significant "negative externalities," such as oceanic dead zones and a loss of agricultural diversity. If future technologies enable us to sustain a population of 9.6 billion by 2050 (the United Nations' estimate), they will no doubt need to be immensely powerful. As such, given that unintended consequences are inevitable byproducts of designed artifacts, we should expect them to give rise to a number of monsters whose deleterious effects are equal in intensity (as we'll discuss again momentarily). The resulting situation could be even *more* dire and precarious.

12. Revolt.
I don't have much to say about how this could work, given the complexities of the contemporary world. Perhaps we'll have to turn to comedian and activist Russell Brand for an answer! The point is that if all else fails—i.e., if incremental *reform* is unable to keep us below critical thresholds and ensure that dangerous technologies are used responsibly—then we may need a clean break in the form of a *revolution*. While I'm not inspired by anarcho-primitivism (although this wasn't always the case), it may turn out to be the most plausible option via a process of elimination.*

Top Priorities

There are numerous big-picture hazards confronting us today and in the near future, and many more should be expected if the development of increasingly powerful dual-use technologies continues. So, which ought to concern us the most? Which should we focus on? How ought we allocate our limited resources?

It's worth noting that the most threatening *existential risks* might not be the most threatening *global catastrophes*. There might be, for example,

---

* I have left off a number of additional possibilities, such as supporting politicians who advocate for nuclear disarmament, creating "a framework for international action," and retaining "a last-resort readiness for preemptive action" (see Nick Bostrom, "Existential Risks: Analyzing Human Extinction Scenarios and Related Hazards," **Journal of Evolution and Technology** 9, no. 1 (2002). Although I don't include them in the list, they ought to be noted.

risks that have all-or-nothing consequences: either they don't affect us at all, *or* they cause profound devastation. Superintelligence appears to be a risk of this sort: if it can destroy half of humanity, it can probably destroy all of it. A superintelligence takeover might thus rank low on a list of sub-existential dangers and high on a list of existential risks. On the other hand, global warming isn't seen by many experts as an urgent existential threat, although nearly *everyone* agrees that its effects could (indeed, will) be catastrophic. It would thus rank low on a list of existential risks and high on a list of large-scale disasters.

To illustrate this idea more concretely, consider the probability rankings of a 2008 conference on big-picture risks hosted by the Future of Humanity Institute. (This conference was mentioned earlier: it placed the overall probability of existential annihilation at 19%.) Table B orders the risks according to their likelihood of causing our extinction (the red-dot scenarios). According to conference participants, the most significant existential risks are: nanotech terror, superintelligence, wars (in general), engineered pandemic, nuclear war, nanotech error, natural pandemic, and nuclear terrorism. But if one were to order these according to the probability of killing "at least 1 million" people, for example, the ordering would be quite different. It would go: wars (in general), natural pandemic, engineered pandemic, nuclear war, nanotech terror, nuclear terrorism, superintelligence, and nanotech error. Here superintelligence is among the least significant threats.

This leads to a troubling question: how do we spend our time, money, and other finite resources? The question would be easy to answer if the "most pressing existential risks" and "most pressing global catastrophes" lists were identical, but they aren't. Consequently, there could emerge *competing priorities* between different kinds of big-picture hazards. If, for instance, we take the existential threat of superintelligence seriously, and therefore funnel our resources into projects working toward the creation of Friendly AI, this could potentially take away from research on, say, vaccines to combat deadly pathogens. The result is that *reducing the probability of an existential catastrophe could actually increase our vulnerability to a variety of nonexistential disasters*, and vice versa. It follows that suffering a series of global catastrophes may be the unavoidable cost of ensuring our continued survival on Earth, and going extinct may be the cost of avoiding a series of global catastrophes. Perhaps we can't have it both ways.

Table B. The Result of an "Informal Survey" at the 2008 Global Catastrophic Risks Conference in Oxford (by percent chance)[54]

RISK	AT LEAST 1 MILLION DEAD	AT LEAST 1 BILLION DEAD	EXTINCTION
Total killed by molecular nanotech	25	10	5
Total killed by Superintelligent AI	10	5	5
Total killed in all wars (including civil wars)	98	30	4
Total killed in the single biggest engineered pandemic	30	10	2
Total killed in all nuclear wars	30	10	1
Total killed in the single biggest nanotech accident	5	1	0.5
Total killed in the single biggest natural pandemic	60	5	0.05
Total killed in all acts of nuclear terrorism	15	1	0.03
Overall risk of extinction prior to 2100	n/a	n/a	19

As it happens, the aforementioned conference survey isn't the only attempt to list the most pressing existential risks in order. The Oxford neuroscientist Anders Sandberg published an article on *The Conversation* in which he identifies the top five risks facing humanity in the following order (from most pressing to least): nuclear war, a bioengineered pandemic, superintelligence, nanotechnology, and unknown unknowns. This is similar in some respects with the conference rankings above. And while people like Stephen Hawking, Elon Musk, and Nick Bostrom haven't compiled "biggest risks" lists *per se*, all have identified (either explicitly or implicitly) superintelligence as probably the greatest threat facing humanity in the coming centuries.

My own list of top existential risks differs from those so far offered:

1. Monsters

The number one threat facing us in the coming centuries is an unknown unknown. This is consistent with the assumption made by both the bottleneck and parallel growth hypotheses, namely that the number of existential risk scenarios will continue to grow in the future, as increasingly powerful dual-use technologies are developed. But the category of monsters is much broader than this. It also encompasses unintended consequences of purposive action, natural phenomena that we don't yet understand, and unforeseen synergies between different risk scenarios unfolding in simultaneity. While some of these unknowns are ultimately knowable, others may be unknowable *in principle*: we may be cognitively closed to them. The lesson here is twofold: (i) we should accelerate the advancement of science as much as possible to convert as many *knowable unknowns* into *knowable knowns*, and (ii) whatever our prior probability estimates of human annihilation are, we should increase them, since there could be a potentially large number of risks that we'll just never see coming, due to inherent limitations in our cognitive machinery. This is, as suggested, the primary reason I think Rees' coin-toss estimate of human extinction this century might be, if anything, on the sunny side.

As the Global Challenges Foundation's landmark publication on big-picture hazards notes, some future risks might "sound unlikely and for many possibly ridiculous. But many of today's risks would have sounded ridiculous to people from the past. If this trend is extrapolated, there will be risks in the future that sound ridiculous today, which means that absurdity is not a useful guide to risk intensity."[55] Ultimately, the question will be not whether we can prevent new monsters from arising, but whether we can effectively prevent them from being realized through error or terror.

2. Superintelligence

I agree with Hawking, Bostrom, and others. This is one of the most formidable risks of the future. It's also one of the most speculative: indeed, we are speculating not only about the *behavior* of a future cognitive superbeing but also the *technology* upon which that superbeing's existence is dependent. The speculativeness of this scenario can be acknowledged without undercutting its genuine significance as a risk.

From a theoretical perspective, it appears entirely possible that a recursively self-improving intelligence (a Seed AI) could emerge from the circuits of a supercomputer quite rapidly. A fast takeoff scenario, Bostrom argues, is more likely than a slow one.[56] If the resulting mind were to prefer enmity over amity, we'd be in big trouble: like an ant trying to defend itself against someone wielding a broom. The same outcome could materialize if the superintelligence is indifferent to our survival. As Yudkowsky puts it, perhaps "the AI does not hate you, nor does it love you, but you are made out of atoms which can use for something else."[57] And even if the superintelligence were to care about humanity, there's still an error option: it could end up killing us purely by accident, as the result of some "freak error" or "stupid mistake." (Recall the orthogonality thesis of fallibility from chapter 5.) In addition, superintelligent beings, whether artificial or not, could be used by governments, groups, or individuals to inflict harm upon others—they could, in other words, be *weaponized* to gain a strategic advantage in the relentless struggle for political domination.

3. Nanotechnology
This is one of two emerging technologies that will become more accessible as it becomes more powerful. Consequently, more and more people will be able to employ it to rearrange the world as they desire. As mentioned several times, the problem with accessibility is that there are far more deranged individuals than malicious groups, and far more malicious groups than rogue governments. So the threat from advanced nanotechnology will become increasingly significant as the field matures. At the extreme, a psychopath with a death wish for humanity could destroy the entire biosphere without us having known he or she was even a threat.

4. Bioterrorism/Engineered Pandemic
This is the other emerging technology where accessibility and power are following parallel exponential trajectories. While highly virulent pathogens tend not to survive in nature, one could exploit synthetic biology and genetic engineering to design microbes for the specific purpose of killing as many people as possible. The terrorist of the future will be an outlaw who knows what "polymerase chain reaction" means, who knows how

to search online genomic databases, and who has an extra few thousand dollars to set up a small laboratory in his or her apartment. The terrorist of the future won't need to be particularly bright, knowledgeable, skillful, or wealthy to wreak unprecedented havoc on human civilization.

5. Nuclear Weapons

A new Cold War between the US and Russia may be heating up, terrorist groups like al-Qaeda and the Islamic State see it as their "religious duty" to obtain nuclear devices, and many scholars assign a shockingly high probability to a nuclear bomb being detonated in a major urban area on a timescale of decades. If even a single bomb were to go off, the result could be millions dead, potentially severe overreactions from governments, worldwide psychological trauma, and firestorms leading to a nuclear winter. This could cause severe crop failures, starvation, malnutrition, social and economic collapse, and—at the extreme—an existential disaster. We have luck to thank for no nuclear devices being used as weapons since Nagasaki, but we can't rely on luck to keep us safe in the future.

6. Biodiversity Loss/Global Warming

We're at the beginning of the sixth mass extinction event in life's 3.5 billion-year history. Organisms are dying out at an extraordinary rate. Between 1970 and 2010, for example, the global population of vertebrates declined by an unbelievable 52%. (Feel free to extrapolate this trend into the future.) And a recent article in *Nature* argues that, because of human activity, we may be on the verge of an irreversible, catastrophic collapse of the global ecosystem.

One of the primary causes of biodiversity loss is global warming. It's noteworthy that significant doubt remains about whether global warming is actually occurring and about whether it's anthropogenic, *but not among the experts*. For those who study the climate, global warming is a clear and present danger. Its effects will be "severe," "pervasive," and "irreversible." In the worst-case scenario, the earth could turn into an unlivable hellhole like Venus, whose climate succumbed to a *runaway greenhouse effect*. It remains an open question how probable this possibility is.

7. Natural Pandemic

The 1918 Spanish flu pandemic sickened about *one-third* of the global population and killed up to 100 million people; the HIV pandemic has infected nearly 78 million people and killed about 39 million; smallpox in the twentieth century killed up to 500 million; the Black Death killed some 75 million; and the 2009 swine flu pandemic affected 74 countries and killed "unquestionably higher" than 18,000 people.[58] Historically speaking, most of the biggest catastrophes didn't involve wars, totalitarian regimes, asteroid impacts, volcanic eruptions, or economic recessions; they involved pathogenic bugs leaping from one sick body to the next. While we can't extrapolate infectious outbreaks from history the way we can with other risks, it's reasonable to say that a future pandemic cooked up in the kitchen of nature could have a significant impact on civilization.

8. Simulation Shutdown

This is perhaps the most speculative possibility we've considered. Until there's a flaw pointed out in the argument, though, we shouldn't dismiss it as hogwash. This being said, if we are in a simulation, and if there exists a number of simulations above us, the threat of a catastrophe could be immense, since annihilation is inherited downward in simulation stacks. Thus, if there are 100 universes above us, any one of these getting shut down—either on accident or on purpose—would result in our own universe suddenly vanishing into the digital oblivion. In the case of humanity running large numbers of simulations in the future, there may be reason for moving this risk to the very top of the list.

9. Supervolcanoes

The effects of a volcanic winter are similar to that of a nuclear winter: crop failure, malnutrition, disease, and economic collapse. We know that supereruptions capable of inducing a climatic disaster occur on average every 50,000 years.[59] Mount Tambora erupted in 1816 and had global consequences, but the Toba eruption "ejected about 300 times more volcanic ash."[60] If the earth were to split open and spew its innards high up into the atmosphere, the future of our species could be seriously jeopardized.

10. Asteroid/Comet Impacts

This is a completely avoidable catastrophe, yet one that we remain vulnerable to at the moment. The effects of a sufficiently large impactor would be more or less identical to that of a supereruption (and therefore a nuclear winter). Although the probability is low, it would be a cosmic embarrassment for a species with our level of technological prowess to kick the bucket from a random collision with a celestial object.[61]

11. Entropy Death

Although not an urgent risk (not even close!), this one appears to be unavoidable. Far into the future, the universe will inexorably sink into a lifeless state of frozen chaos. But long before this happens, the sun will sterilize the earth as it turns into a red giant and eventually swallow it whole.

As with the other lists, this ordering cannot be taken too literally. Our visibility into the future remains, in important respects, rather low. For example, there's no overwhelmingly cogent argument for placing nanotechnology above an engineered pandemic (although there *is* reason for placing it above an asteroid impact). It may be that as we proceed forward in time, rankings like those above will converge. Understanding what lies ahead is like driving on a dark road at night: as you approach the objects before you, they become increasingly discernible, until you can finally identify them with certainty. Perhaps as the twenty-first century unfolds and our brash adventure into the wilderness of risk continues, it will become increasingly clear what the most pressing hazards are to our collective survival and prosperity. Or perhaps new monsters will perpetually obfuscate our view.

The End Is Here

As you read this sentence, the earth sits at the center of a giant, expanding bubble of leaked electromagnetic radiation hurtling through the universe at the cosmic speed limit of light. The outermost edges of this bubble—now some 140 light-years in diameter—contain programs like *The Howdy Doody Show*, *Meet the Press*, and *I Love Lucy* that were broadcast by early television stations, a major source of leakage, along with military radar.[62] These signals were picked up by the antenna atop people's homes, but some ricocheted into the vastness of space.

A civilization living on the far side of our galaxy could, with the right technology (namely radio telescopes), potentially detect these stray signals from Earth and infer the existence of another life-form—one clever enough to accidentally seep streams of very high frequency (VHF) light into the sky. (Unfortunately, the inverse square law entails a significant degradation of these signals as they travel further from Earth.) The reverse situation applies as well: if a civilization developed to our level of sophistication, we could potentially detect its bubble of radiation spillage expanding toward our lonely planet.

Yet in all directions, the universe looks like a barren wasteland of dead matter mindlessly acting out a cosmic screenplay written by nature's laws. This panoply can be beautiful, for sure, but it lacks any convincing signs that intelligent life is crying out for companionship in a universe bereft of intrinsic meaning. There are no ripples of leaked radiation splashing against the shores of Earth. The sky is quiet—not a whisper, much less a shriek. The conundrum is that this is exactly opposite of what we would expect, given what we know about the natural world. The universe should be teeming with life, according to some estimates using the Drake Equation; we should be able to point our telescopes at the midnight firmament and see, at least on occasion, a spaceship flying by.

What explains this Great Silence? Perhaps there is a period in every intelligent species' life at which the *archaic* (old brains filled with old ideas about the universe) catastrophically collides with the *neoteric* (new technologies capable of rearranging the universe in new ways). We live in a world, today, in which scientists are literally studying the first *billionth* of a second after the Big Bang. At the very same time, a sizable portion of evangelicals await being suddenly "caught up" into the clouds with Jesus, and many Muslims are eagerly anticipating the appearance of the Mahdi, followed by Jesus in the arms of two angels. We live in a world that contains both Noam Chomsky and John Hagee, Sir Martin Rees and Sarah Palin, and Steven Pinker and Abu Bakr al-Baghdadi. If the rationality of our *ends* fails to match the rationality of our *means*, there's a fairly good chance that this century, the forty-five millionth since Earth formed, could be our last.

The universe, as J. B. S. Haldane once quipped, is not only stranger than we imagined, but stranger than we *could have* imagined. Consider that matter is literally more than 99.9% empty space; one species can morph into another species through evolution; the loudest animal relative to its body size is a tiny insect that produces sound by rubbing its penis against its stomach; the fastest organism in the world is the white mulberry tree (which propels its pollen at half the speed of sound); the universe literally has no center and no boundaries; fruit flies have sperm that are 2.3 inches long when uncoiled; as part of its life cycle, a marine invertebrate called the tunicate eats its own brain; the atmosphere gets colder, then warmer, then colder as you move toward space; most humans have some Neanderthal genes; the cosmos isn't expanding *into* space, rather space *itself* is expanding; if you leave your refrigerator door open on a hot summer day, the room will warm up rather than cool down; if you stop suddenly with a helium balloon in your car, it will dart backward rather than forward; and if you add up every single number from one to infinity, the resulting sum will be exactly *negative one-twelfth*.[63]

Who knows what other mysteries await our discovery. Who knows what wonders might dazzle our imaginations and tickle our intellects with eureka, "ah-ha" moments. And who knows what our frail-brained species could become if the *good* uses of technology are allowed to realize the evolutionary adventure of redesigning our bodies and brains for purposes that conduce to happiness and morality.

While science, philosophy, art, culture, music, literature, poetry, fashion, sports, and all the other objects of civilization make life *worth* living, avoiding an existential catastrophe makes it *possible*. This makes eschatology, with its two interacting branches, the most important subject that one could study. Without an understanding of what the risks are before us, without an understanding of how the clash of eschatologies has shaped the course of world history, we will be impotent to defend against the threat of (self-)annihilation. The fact is that we are the only remaining species of human on the planet—the last one, *Homo floresiensis*, having died out some 12,000 years ago in Indonesia. Our situation has always been precarious, but it's never been as precarious as it is today. If we want our children to have the opportunity of living the Good Life, or even existing at all, it's essential that we learn to favor evidence over faith, observation over revelation, and science over religion as we venture into a dangerously wonderful future.

Notes

1.   See Nick Bostrom, "Existential Risk Prevention as Global Priority," *Global Policy* 4, no. 1 (2013), http://www.existential-risk.org/concept.pdf.

2.   See "The Russell-Einstein Manifesto," July 9, 1955, http://pugwash.org/1955/07/09/statement-manifesto/. Italics added.

3.   After mentioning this passage in my first book, *A Crisis of Faith* (Dangerous Little Books, 2012), I came across it quoted in a very similar context in Sir Martin Rees' *Our Final Hour* (Basic Books, 2003). This is purely coincidental, although it behooves me to mention the parallelism.

250 • THE END

4. See "STERN REVIEW: The Economics of Climate Change," accessed June 18, 2015, http://www.wwf.se/source.php/1169157/Stern%20Report_Exec%20Summary.pdf.

5. See Anders Sandberg and Nick Bostrom, "Global Catastrophic Risks Survey," Future of Humanity Institute, 2008, http://www.fhi.ox.ac.uk/gcr-report.pdf.

6. Nick Bostrom, "Existential Risks: Analyzing Human Extinction Scenarios and Related Hazards," *Journal of Evolution and Technology* 9, no. 1 (2002).

7. See Sir Martin Rees, *Our Final Hour: A Scientist's Warning* (Basic Books, 2003).

8. See Langdon Winner, *Autonomous Technology: Technics-out-of-Control as a Theme in Political Thought* (MIT Press, 1978), p. 97. See also the classic work by Jacques Ellul, *The Technological Society* (Alfred A. Knopf, Inc., 1964).

9. See Bill Joy, "Why the Future Doesn't Need Us," *Wired*, April 2000, http://archive.wired.com/wired/archive/8.04/joy.html.

10. This section draws heavily, and in some cases *ad verbum*, from Phil Torres, "Are We Passing Through a Bottleneck, or Will the Explosion of Existential Risks Continue?" IEET, January 25, 2015, http://ieet.org/index.php/IEET/more/torres20150125.

11. See Sir Martin Rees, *Our Final Hour: A Scientist's Warning* (Basic Books, 2003), p. 8.

12. See Nick Bostrom, "The Future of Humanity," in *New Waves in Philosophy of Technology*, ed. Jan-Kyrre Berg Olsen, Evan Selinger, and Soren Riis (Palgrave McMillan, 2009), pp. 186–216.

13. See Nick Bostrom, "Existential Risk Prevention as Global Priority," *Global Policy* 4, no. 1 (2013), http://www.existential-risk.org/concept.pdf.

14. See Nick Bostrom, "Letter from Utopia," *Studies in Ethics, Law, and Technology* 2, no. 1 (2008), http://www.nickbostrom.com/utopia.html.

15. The definition of "Singularity" here comes from Ray Kurzweil, "Law of Accelerating Returns," 2001, http://lifeboat.com/ex/law.of.accelerating.returns. Notice that "singularity" is used in multiple ways. Sometimes it refers to the intelligence explosion discussed in chapter 5; other times it refers to the increasingly rapid pace of technological development more generally.

16. See Ray Kurzweil, *The Singularity Is Near* (Penguin Group, 2005).

17. Again, this is a reference to Bostrom's "Maxipok" rule of thumb. See Nick Bostrom, "Existential Risk Prevention as Global Priority," *Global Policy* 4, no. 1 (2013), http://www.existential-risk.org/concept.pdf.

18.   In his fascinating book *Mind Children*, the AI pioneer and transhumanist Hans Moravec suggests (to paraphrase John Leslie's summary of him) that the human species is in its final century of existence, and that we should work hard for its speedy extinction. See John Leslie, "The Risk That Humans Will Soon Be Extinct," *Philosophy* 85, no. 4 (2010), and Hans Moravec, *Mind Children: The future of Robot and Human Intelligence* (Harvard University Press, 1988).

19.   As the philosopher Mark Walker colorfully puts it, "The upper bound on what we may hope for might be that our transcendent children will fulfill the prophecy of the second coming with only a small inversion of the etiology: It is not God that creates a man-god but we humans create god-like beings." See Mark Walker, "Prolegomena to Any Future Philosophy," *Journal of Evolution and Technology* 10 (2002), http://www.jetpress.org/volume10/prolegomena.html.

20.   See Nick Bostrom, "The Future of Humanity," in *New Waves in Philosophy of Technology*, ed. Jan-Kyrre Berg Olsen, Evan Selinger, and Soren Riis (Palgrave McMillan, 2009), pp. 186–216.

21.   See Ray Kurzweil, *The Singularity Is Near* (Penguin Group, 2005).

22.   See Ed White, "NASA Recruiting Volunteers for 'Out of This World' Jobs," Air Force Space Command, Februrary 6, 2008, http://www.afspc.af.mil/news/story.asp?id=123085237.

23.   See Roger Highfield, "Colonies in Space May Be Only Hope, Says Hawking," *Telegraph*, October 16, 2001, http://www.telegraph.co.uk/news/uknews/1359562/Colonies-in-space-may-be-only-hope-says-Hawking.html.

24.   See Jacqueline Howard, "Elon Musk Wants To Put a Million People on Mars," *Huffington Post*, September 30, 2014, http://www.huffingtonpost.com/2014/09/30/elon-musk-mars-aeon-interview_n_5907914.html.

25.   See Chris Matyszczyk, "Stephen Hawking: God Particle Could Wipe Out Universe," *CNET*, September 7, 2014, http://www.cnet.com/news/stephen-hawking-god-particle-may-wipe-out-the-universe/.

26.   For more, see Jason Matheny, "Reducing the Risk of Human Extinction," *Risk Analysis* 27, no. 5 (2007): p. 1337, and Anders Sandberg, Jason Matheny, and Milan Ćirković, "How Can We Reduce the Risk of Human Extinction?" *Bulletin of the Atomic Scientists*, September 9, 2008, http://thebulletin.org/how-can-we-reduce-risk-human-extinction.

27.   See Nick Bostrom, "Existential Risks: Analyzing Human Extinction Scenarios and Related Hazards," *Journal of Evolution and Technology* 9, no. 1 (2002).

28.   See Nick Bostrom, *Superintelligence: Paths, Dangers, Strategies* (Oxford University Press, 2014).

29.   See Enrico Moretti, "Does Education Reduce Participation in Criminal Activities?" (lecture, symposium on the "Social Costs of Inadequate Education," Columbia University, October 25–25, 2005), https://www.tc.columbia.edu/centers/EquitySymposium/symposium/resourceDetails.asp?PresId=6.

30.   See Peter Bergen and Swati Pandey, "The Madrassa Myth," *New York Times*, August 2005, http://www.nytimes.com/2005/06/14/opinion/the-madrassa-myth.html?_r=0.

31.   See Aaron Y. Zelin, "Abu Bakr al-Baghdadi: Islamic State's Driving Force," *BBC*, July 31, 2014, http://www.bbc.com/news/world-middle-east-28560449.

32.   See Aaron Y. Zelin, "Abu Bakr al-Baghdadi: Islamic State's Driving Force," *BBC*, July 31, 2014, http://www.bbc.com/news/world-middle-east-28560449.

33.   See Peter Boghossian, *A Manual for Creating Atheists* (Pitchstone Publishing, 2013).

34.   See Will Gervais and Ara Norenzayan, "Analytic Thinking Promotes Religious Disbelief," *Science* 336 (2012): 6080, http://www.sciencemag.org/content/336/6080/493.

35.   See Valerie Strauss, "Texas GOP Rejects 'Critical Thinking' Skills. Really." *Washington Post*, July 9, 2012, http://www.washingtonpost.com/blogs/answer-sheet/post/texas-gop-rejects-critical-thinking-skills-really/2012/07/08/gJQAHNpFXW_blog.html.

36.   And no studies show the opposite. See Chris Mooney, "The Science of Fox News: Why Its Viewers Are the Most Misinformed," *Alternet*, April 8, 2012, http://www.alternet.org/story/154875/the_science_of_fox_news%3A_why_its_viewers_are_the_most_misinformed.

37.   See Michael Kelley, "STUDY: Watching Only Fox News Makes You Less Informed Than Watching No News at All," *Business Insider*, May 22, 2012, http://www.businessinsider.com/study-watching-fox-news-makes-you-less-informed-than-watching-no-news-at-all-2012-5.

38.   See Bruce Bartlett, "How Fox News Changed American Media and Political Dynamics," *Economic Review* (2005), http://www.ashford.zone/images/2015/05/SSRN-id2604679.pdf.

39.   See Naomi Oreskes and Erik Conway, *The Collapse of Western Civilization: A View from the Future* (Columbia University Press, 2014).

40.   See Kenneth Chang, "Bias Persists for Women of Science, a Study Finds," *New York Times*, September 24, 2012, http://www.nytimes.com/2012/09/25/science/bias-persists-against-women-of-science-a-study-says.html?_r=0.

41.   See Nick Bostrom, *Superintelligence: Paths, Dangers, Strategies* (Oxford University Press, 2014), p. 54.

42.   See Anita Williams Woolley et al., "Evidence for a Collective Intelligence Factor in the Performance of Human Groups," *Science* 330 (2010): 6004, http://www.sciencemag.org/content/330/6004/686.abstract;   and   "Collective Intelligence: Number of Women in Group Linked to Effectiveness in Solving Difficult Problems," *ScienceDaily*, October 2, 2010, http://www.sciencedaily.com/releases/2010/09/100930143339.htm.

43.   See "Collective Intelligence: Number of Women in Group Linked to Effectiveness in Solving Difficult Problems," *ScienceDaily*, October 2, 2010, http://www.sciencedaily.com/releases/2010/09/100930143339.htm.

44.   See Robin Hanson, "Catastrophe, Social Collapse, and Human Extinction," in *Global Catastrophic Risks*, ed. Nick Bostrom and Milan Ćirković (Oxford University Press, 2008). For further discussion, see Jason Matheny, "Reducing the Risk of Human Extinction," *Risk Analysis* 27, no. 5 (2007).

45.   See "Banking against Doomsday," *Economist*, March 8, 2012, http://www.economist.com/node/21549931.

46.   See Laura Shin, ""The 85 Richest People in the World Have as Much Wealth As The 3.5 Billion Poorest," *Forbes*, January 23, 2014, http://www.forbes.com/sites/laurashin/2014/01/23/the-85-richest-people-in-the-world-have-as-much-wealth-as-the-3-5-billion-poorest/.

47.   See Anup Shah, "Poverty Facts and Stats," Global Issues, January 7, 2013, http://www.globalissues.org/article/26/poverty-facts-and-stats.

48.   See Asle Ronning, "The Well-Healed Leave Biggest Carbon Footprint," *ScienceNordic*, August 14, 2013, http://sciencenordic.com/well-heeled-leave-biggest-carbon-footprint.

49.   See Liz Minchin, "Carbon Footprint of Rich Twice That of Poor," *Age*, June 16, 2007, http://www.theage.com.au/news/national/carbon-footprint-of-rich-twice-that-of-poor/2007/06/15/1181414549948.html.

50.   See "Ecological Footprinting," Development Education, http://www.developmenteducation.ie/de-in-action/ecological-footprinting/do-we-all.html.

51.   See Robert Winnett, "President George Bush: 'Goodbye from the World's Biggest Polluter," *Telegraph*, July 9, 2008, http://www.telegraph.co.uk/news/

worldnews/2277298/President-George-Bush-Goodbye-from-the-worlds-biggest-polluter.html.

52. See Charles Q. Choi, "Rich People More Likely to Lie, Cheat, Study Suggests," *Live Science*, February 27, 2012, http://www.livescience.com/18683-rich-people-lie-cheat-study.html.

53. See "Living Planet Report 2014," World Wildlife Fund, October 2014, http://wwf.panda.org/about_our_earth/all_publications/living_planet_report/.

54. See Anders Sandberg and Nick Bostrom, "Global Catastrophic Risks Survey," Future of Humanity Institute, 2008, http://www.fhi.ox.ac.uk/gcr-report.pdf.

55. See "12 Risks That Threaten Human Civilisation," Global Challenges Foundation, February 2015, p. 127.

56. See Nick Bostrom, *Superintelligence: Paths, Dangers, Strategies* (Oxford University Press, 2014), p. 65.

57. Eliezer Yudkowsky, "Artificial Intelligence as a Positive and Negative Factor in Global Risk," Machine Intelligence Research Institute, https://intelligence.org/files/AIPosNegFactor.pdf ("This version contains minor changes.")

58. See Simeon Bennett, "Swine Flu Deaths May Have Been 15 Times Higher Than Reported, *Bloomberg Busness*, June 25, 2012, http://www.bloomberg.com/news/articles/2012-06-25/swine-flu-deaths-may-have-been-15-times-higher-than-reported.

59. See Michael Rampino, "Super-volcanism and Other Geophysical Processes of Catastrophic Import," in *Global Catastrophic Risks*, ed. Nick Bostrom and Milan Ćirković (Oxford University Press, 2008), p. 215.

60. See "Super-eruptions: Global Effects and Future Threats," Geological Society of London Working Group, p. 6.

61. See Neil deGrasse Tyson, "We Can Survive Kill Asteroids—But It Won't Be Easy," *Wired*, April 2, 2012, http://www.wired.com/2012/04/opinion-tyson-killer-asteroids/.

62. See Barry Parker, *Alien Life: The Search for Extraterrestrials and Beyond* (Basic Books, 1998), pp. 171–174.

63. While this is not inaccurate, the statement requires some qualifications. For a good discussion of how this infinite series could possibly equal a negative fraction, see Evelyn Lamb, "Does 1+2+3 Really Equal -1/12?" *Scientific American*, January 20, 2014, and Phil Plait, "Follow-Up: The Infinite Series and the Mind-Blowing Result," *Slate*, January 18, 2014.

Acknowledgments

· · · · ·

This book was written during a period of globe-trotting itinerancy. My partner, Whitney Trettien, and I sold our house in Durham, North Carolina, to spend the year living in Airbnb apartments and friends' homes all over Europe and North America. Significant portions of this book were written in Geneva, Lausanne, Ovronnaz, and Morgins (Switzerland); Leipzig and Berlin (Germany); Boca Raton (Florida); San Francisco (California); Durham (North Carolina); Washington, DC; Ithaca (New York); London, Fishguard, and Blaenau Ffestiniog (UK); Dublin (Ireland); Vancouver (Canada); St. Petersburg (Florida); Frederick (Maryland); and finally, Carrboro (North Carolina), where I now live. I have, more than anyone else, my partner to thank for making this laborious project possible. She is a brilliant scholar, perpetual source of inspiration, and truly amazing person. I love her more than words can express (as such clichés make evident).

While the book was written largely in isolation—in busy London cafés or ten-story balconies overlooking Berlin—once I had a rough draft (which turned out to be very rough, as all writing is rewriting), I benefited from incisive comments and criticisms by a significant number of scholars. I am greatly indebted for feedback from (in no particular order) Russell Blackford, Andrew Maynard, Michael Rampino, Alan Hale, Roman Yampolskiy, Barry Dainton, J. M. Berger, Matthias Steup, Adam B. Smith, Michael Svigel, Joseph Siracusa, Daniel Kirk-Davidoff, Marcin Kulczycki, Herb Silverman, Fred Adams, Ron Geaves, Chris Phoenix, Tony Barrett, David Cook, Jean Guillemin, John Loftus, Gerrit Kreyenbroek, Peter Forrest, Stephen Spector, S. Michael Houdmann, Larry Poston, Robert Smith, Owen Cotton-Barratt, Matthew Sharpe, Jønathan Lyons, Bruce Riedel, Kurt Volkan, Fran McDonald, Kris Notaro, William

Chittick, Ken Olum, and Peter Lewis. Special thanks to John Leslie, who provided extensive comments on the entire book, as well as Hank Pellissier, Marcelo Rinesi, Rick Searle, and (especially) Sean Mannion, who provided such extraordinarily detailed comments about spelling, punctuation, and grammar, that it left me in a state of "shock and awe." I'd also like to thank Milan Ćirković for many stimulating emails about existential risk-related issues, and the Institute for Ethics and Emerging Technologies (IEET) for its generous support of my work. In fact, some of the book's chapters, either in part or in whole, were initially posted on the IEET's website as articles. Thanks to all the intelligent IEET readers who commented and critiqued my incipient ideas. Finally, I must add that I take full responsibility for any and all errors that this book may contain, and the opinions presented above are my own.

I would also like to thank my parents, Stephanie Torres and J. P. Torres, my wonderful sister, her husband, and their daughter, Lucy (for being so lovable), and Phyllis Trettien, who hosted us in Florida while my partner flew around the country for job interviews. She's now an assistant professor at the University of North Carolina, Chapel Hill.

Appendix 1

Alternative Risk Typologies

·····

Nick Bostrom's original risk typology (see table A-A) first appeared in his 2002 paper "Existential Risks: Analzing Human Extinction Scenarios and Related Hazards," published in the *Journal of Evolution and Technology*.[1] As one can see, it's a two-dimensional matrix that doesn't account for the temporality of risk scenarios: the y-axis includes three categories that are entirely spatial in nature. Later, Bostrom updated this typology in his co-edited 2008 book with Milan Ćirković called *Global Catastrophic Risks* (see table B-A).[2] This new typology introduced a temporal dimension, but in a way that's both conceptually incoherent and empirically inaccurate.

The conceptual problem is that the categories of the y-axis aren't all of the same sort, which they ought to be if one wants to avoid a category error. There are three spatial categories ("personal," "local," and "global") and then, moving up the y-axis, a spatio*temporal* category ("transgenerational," which essentially means "global and forever") appears. The empirical problem is that there are risks with transgenerational effects that are personal and local in scope, for example, such as germline mutations. Bostrom and Ćirković's

Table A-A. Original Risk Typology

SCOPE			
Global	*Thinning of the ozone layer*	*X*	
Local	*Recession in a country*	*Genocide*	
Personal	*Your car is stolen*	*Death*	
	Endurable	Terminal	INTENSITY

typology implies that only risks with global consequences can have time-extended effects, and this is factually inaccurate.

Similar issues arise with the x-axis. Here, Bostrom and Ćirković define "terminal" as either "causing death *or* permanently and drastically reducing quality of life" (italics added).[3] In other words, some terminal events are actually *endurable*. By analogy, this would be like creating a typology that distinguishes between "dead" and "alive," but then defines "dead" as meaning "either dead *or* alive." Making matters worse, Bostrom and Ćirković further identify "imperceptible" and "endurable" as distinct x-axis categories. But this is analogous to distinguishing between "kitchens" and "houses" in a typology of places to live. The problem is that kitchens aren't an *alternative* to houses, they're a *part* of them. Similarly, imperceptible is a subcategory within the larger category of endurability. It describes a *type* of endurable event, just like the terms "somewhat perceptible," "quite perceptible," and "blatantly obvious." None of these properties are on par with the categories of "endurable" and "terminal," which, if understood in their ordinary, dictionary-defined senses, *do* constitute genuine alternatives.

Table B-A. Updated Risk Typology

SCOPE		Imperceptible	Endurable	Terminal	(Hellish?)
(Cosmic?)					
Trans-generational		*Loss of one species of beetle*	*Drastic loss of biodiversity*	*Human extinction*	
Global		*Global warming by 0.001 degrees*	*Spanish flu pandemic*	*Aging?*	
Local		*Congestion from one extra vehicle*	*Recession in a country*	*Genocide*	
Personal		*Loss of one hair*	*Car is stolen*	*Fatal car crash*	

INTENSITY ⟶

Note: Light grey area represents global catastrophic risks; dark grey area represents existential risks

Table C-A. Third Risk Typology

(Cosmic?)				
Pan-generational	*One original Picasso painting destroyed*	*Destruction of cultural heritage*	*X*	
Trans-generational	*Loss of one species of beetle*	*Global Dark Age*	*Aging*	
Global	*Global warming by 0.01 degrees*	*Thinning of ozone layer*	*Ephemeral global tyranny*	
Local	*Congestion from one extra vehicle*	*Recession in a country*	*Genocide*	
Personal	*Loss of one hair*	*Car is stolen*	*Fatal car crash*	
	Imperceptible	Endurable	Crushing	(Hellish?)

SCOPE (vertical arrow)

SEVERITY ⟶

Note: Light grey area represents global catastrophic risks; dark grey area represents existential risks

Bostrom later proposed another typology (see table C-A) in a 2012 *Global Policy* paper titled "Existential Risk Prevention as Global Priority."[4] Unfortunately, this typology also conflates the temporal and spatial scopes of a risk's consequences, and it fails to fix the conceptual problems of the x-axis, even though it does replace "terminal" with the slightly more colorful term "crushing." (Note that the green squiggly lines are included in the image itself.) These aren't trivial matters: an adequate account of risks ought to be conceptually coherent and empirically accurate. The typologies above and below aren't.

The ultimate thrust of my critique here is the following: the definition of "existential risks" that I adopt throughout this book is nearly identical to the definition given by Bostrom (and Ćirković), namely a catastrophe that either annihilates us or results in the permanent and drastic curtailing of our future prospects. My problem isn't with the destination, as it were, but the journey leading to this destination. I recommend discarding all these typologies and adopting figure B, found in chapter 1, because it overcomes the problems here discussed. It also makes explicit the two necessary conditions that a risk must satisfy to count as existential, which none of Bostrom's typologies do. If one were to glance at such typologies, one could easily come away with the wrong impression: for example, that existential risks are necessarily *terminal* events, which simply isn't the case. An existential risk could be terminal *or* endurable, as figure B makes clear.

Notes

1. See Nick Bostrom, "Existential Risks: Analyzing Human Extinction Scenarios and Related Hazards," *Journal of Evolution and Technology* 9, no. 1 (2002).

2. See Nick Bostrom and Milan Ćirković, "Introduction," in *Global Catastrophic Risks*, ed. Nick Bostrom and Milan Ćirković (Oxford University Press, 2008).

3. See Nick Bostrom and Milan Ćirković, "Introduction," in *Global Catastrophic Risks*, ed. Nick Bostrom and Milan Ćirković (Oxford University Press, 2008), p. 4.

4. See Nick Bostrom, "Existential Risk Prevention as Global Priority," *Global Policy* 4, no. 1 (2013), http://www.existential-risk.org/concept.pdf.

Appendix 2

The Story of Zoroastrianism

• • • • •

Thus Spoke Zoroaster

For most of recorded history, eschatology has been thoroughly religious in nature. Its narratives have focused on miraculous events involving supernatural agents fighting a cosmic battle between the moral opposites of Good and Evil. As mentioned in chapter 1, one of the earliest specimens of systematic eschatological thinking comes from the ancient Persian prophet Zoroaster (also called Zarathustra), perhaps 3,000 years ago. The Zoroastrian narrative is a marvelously colorful collage of now-familiar themes, including a virgin-born "Savior" (the *Saoshyant*) who ushers in a Resurrection of the Dead, an Armageddon-like battle between God (Ahura Mazdâ) and Satan (Ahriman), and a Final Judgment of all human souls amid a grand transformation of the cosmos. Such parallelisms have led many scholars to conclude that Zoroastrian eschatology probably influenced the eschatology of Judaism, and therefore the eschatology of Christianity and Islam. If this were the case, then *Zoroaster, the historical figure, might very well be the most influential person in all of human history, having affected world culture, from East to West, more than Confucius, Plato, Jesus, Muhammad, and even Karl Marx.* Thus spoke Zoroaster, and then everyone else plagiarized him.

Let's begin this section of appendix 2 with the Christmas story. We all know how it goes: Mary and Joseph traveled to Bethlehem for a worldwide census (the Census of Quirinius), and while taking shelter in a stable because the inn was full, Mary gave birth to Jesus, the Son of God. Shortly afterward, several "wise men," or *Magi*, from the East followed the brilliant Star of Bethlehem to the stable. Bowing before Jesus, they worshipped him and offered gifts of gold, frankincense, and myrrh.

But what do we know of these strange travelers? Who were the Magi that visited Jesus? It's quite possible that they were adherents of Zoroastrianism, since "Magi" specifically denotes members of the Zoroastrian priesthood. (The term appears in the oldest Greek texts.) If so, it may have been that these Magi weren't actually looking for Jesus but the messianic figure prophesied by their own religion, the last of Zoroaster's three virgin-born sons, the *Saoshyant*. For the three wise men wandering the Middle East long ago, their journey may have had a completely different purpose than what Christians today superimpose upon it. Thus, next time you see a Christmas card with these globetrotters next to Jesus, think of Zoroaster!

The Beginning and End of Time (Literally)

Like all great prophets, Zoroaster obtained his "knowledge" about the ultimate nature and purpose of the universe through divine revelation. According to his revelation, the world exists because Ahriman was lurking about, and it occurred to Ahura Mazdâ that the only way to defeat Ahriman was to build a universe in which good and evil could go to war. Ahura Mazdâ determined that this universe should last for exactly 12,000 years and consist of exactly 4 distinct epochs (one might call them "dispensations").* Zoroaster himself appears at the beginning of the fourth epoch, thereby ushering in the final tri-millennium of cosmic history. Since Zoroaster may have been born around 1,000 BC, it follows that we're probably living in the Zoroastrian end times right now.

The narrative of cosmic history, according to Zoroaster, goes as follows: the initial 3,000-year period consists of spiritual creation only. At this point, the world takes on an "ideal," nonmaterial state, and time does not yet exist. (It's not clear how this could be, of course, since years are a measurement of time.) Toward the end of this epoch, Ahura Mazdâ offers Ahriman a chance to coexist with him in peace, but Ahriman foolishly refuses. Ahura Mazdâ then cuts out a piece of eternal time "during which the battle between good and evil would be fought" and recites a sacred hymn (still uttered by Zoroastrians today).[1] The power of this glorious hymn causes Ahriman to fall into darkness for the next 3,000-year period.

---

* Technically the world's history is 12,057 years long, as we'll examine below, since the Saoshyant appears at year 12,000. Note that an alternative account has it that history consists of three epochs and lasts a total of 9,000 years.

During the second epoch, Ahura Mazdâ converts the world from a nonmaterial to a material state, albeit one that's still "ideal." The cosmos as a whole takes the shape of a great sphere that envelopes the earth, which is flat and floats upon a large ocean. The spherical heavens above are made to revolve around a single mountain in the middle of our terrestrial island, called Mount Harîtî. Ahura Mazdâ also creates time, although he doesn't yet enable it to progress forward, and consequently the sun remains fixed directly overhead for the duration of this epoch. There are 7 distinct types of creation in the world, according to Zoroastrians: humanity, animals, plants, metal, water, fire, and the earth. All these types are made real in this epoch except for fire, which is associated with change and movement. Thus, we have the earth, which contains metals and floats upon water, and on top of the earth stands a single man, bull, and plant. This is the extent of Ahura Mazdâ's creation at its second stage of becoming: three living creatures paralyzed in a world of stasis, grounded on a buoyant disk enclosed in a sphere. The end of the second epoch marks a pivotal transition in cosmic history. For the first time, Ahura Mazdâ's creation becomes animated, as the dynamic element of fire is introduced. Ahura Mazdâ also invites the eternal component of the human soul to enter the world and help him fight Ahriman, and it agrees.[2] (We have three parts to our souls, only one of which is eternal.) Humanity is now ready to help Ahura Mazdâ and his divine servants trample evil.

But Ahriman emerges from his slumber-like state and lashes out almost immediately, devastating the goodness of Ahura Mazdâ's world: he creates darkness to spoil light, turns sweet water briny, and makes smoke a product of fire.[3] Not only does Ahriman fill the cosmic sphere with noxious pollution, but he murders the first human, bull, and plant. Just as history crosses its halfway point and the world of *Mixture*—in which good and evil first dynamically interact—begins, Ahriman appears to have won. But his victory is illusory, as semen from the first human and bull are spilled into the world before their deaths. The human's fluids give rise to the rest of humanity, while the bull's bring forth the multitude of animal species alive today.[4] As a result, goodness bounces back from Ahriman's attack, and the world transforms into a tumultuous theater of constant warring. This is why, according to Zoroastrianism, we have day and night, docile and dangerous animals, plains and mountains, and so on.[5] The world of Mixture includes both the third and fourth epochs, and won't reach its culmination until evil is finally extirpated and the universe is returned to an ideal state of pure immateriality.

The rest of the third epoch's narrative is a complicated tangle of legendary battles and mythic figures. A long series of violent acts involving ancient heroes and foes continues up to the very end of the tri-millennium, at which point Zoroaster is born. According to one text, Zoroaster's "pre-soul" was fashioned by Ahura Mazdâ at the beginning of the third epoch, when Ahriman nearly emerged the victor from his sudden onslaught. This pre-soul was delivered via a particular type of plant, described as being golden-green, fragrant, and tall; the body of Zoroaster came from rain that fell on the plant. Zoroaster's mother then mixed this plant with milk and drank the resulting concoction, later giving birth to the great prophet and founder of Zoroastrianism.[6]

We've now ventured into the final 3,000-year period of cosmic history. Zoroaster received his revelation from Ahura Mazdâ at age 30 and spent his life battling evil.[7] He is eventually murdered in the ancient Afghan city of Balkh—a center of Zoroastrianism—after a group of Persians known as Turanians attacked the city. Nonetheless, Zoroaster's semen is preserved in a body of water called Lake Kayânsê.

In the final two millennia of this epoch, three virgins swim in this lake and get pregnant from Zoroaster's sperm. These women then give birth to Zoroaster's three savior-sons, the "life-givers" or "revitalizers."[8] Each is born a millennium apart, the first in the year 10,000, the second in 11,000, and the third at the very end of time, in 12,000. (The last of Zoroaster's sons lives 57 years, which is why cosmic history is 12,057 years in total.) Like his father, the first *saoshyant* receives a revelation from Ahura Mazdâ for humanity. A number of miraculous events occur, such as the sun standing perfectly still above the earth for 10 days and nights.[9] The conditions of life gradually improve: trees remain green for years at a time and "the effectiveness of medicines [are] such that none will die of illness."[10] Unfortunately, the close of this millennium sees a reversal of this trend: an evil sorcerer—related to the man who murdered Zoroaster—causes a number of destructive events, and there is a three-year-long winter, bitter and cold, after a period of torrential rains.[11]

But the momentum changes again, as Zoroaster's second savior-son is born in 11,000, inaugurating the very last millennium of world history. As was the case with his older brother, Ahura Mazdâ gives this son a revelation, and a number of miracles take place. The sun remains fixed at the zenith this time for 20 days and nights,[12] and trees remain verdant for six straight years. Life improves after the sorcerer's attack, but as with the previous millennium, these improvements are not sustained.[13]

By the end of this period, nearly 12,000 years after Ahura Mazdâ's original creation, the conditions of life deteriorate. It's here that Zoroaster's third son makes his prophetic appearance, bringing about a number of significant eschatological events. For example, he oversees the Resurrection of the Dead, which is followed by a Final Judgment whereby the contributions each man and woman made during their lives are meticulously assessed. Some sinful individuals will be sent to hell, where they will undergo torture for three days and nights.[14]

After this a great dragon swoops down from the celestial realm and sets fire to the earth. Recall that one of the basic types of creation is metal. Thus, engulfed by fire, the metal within Earth's wrinkles becomes molten and viscously seeps down into the valleys. This forms a behemoth river, which humanity must cross in order to be fully cleansed.[15] Those individuals free of sin will pass through its currents without distress; it "will be like a bath in warm milk."[16] But those who still carry around corrosive thoughts, words, and deeds, will experience tremendous pain while submerged in liquid metal. This is, of course, intriguingly similar to the Islamic prophecy that all humans will be forced to cross a bridge over hell, as some scholars have noted.

Nonetheless, at the end of this process all of humanity will have been purged of their sins, and "all people [will] come together in great love for one another. Fathers, sons, brothers, all men who were friends, ask other men: Where were you all those years, and what judgment did your soul receive?"[17] This leads to a final battle between good and evil, one in which Ahriman "will be greatly and exceedingly smashed by the magic power of the *Gathas*," a collection of 17 hymns composed by Zoroaster. As a result, Ahriman will fall "back to the darkness and gloom through the passage through the sky through which they first rushed in" and the dragon "will be burnt by that molten metal," which "will flow into hell, and that stench and filth in the earth where hell was will be burnt by that metal and become pure." Finally, all the mountains of Earth, which are seen as corruptions of flat land, will be razed.[18]

With the permanent elimination of evil, death, and all other traces of turpitude, the world will enter into an eternal state of absolute perfection (one in which time itself may cease to be[19]), and the book of cosmic history will close.[20]

Notes

1. See Prods Oktor Skjærvø, "Introduction to Zoroastrianism," http://www.fas.harvard.edu/~iranian/Zoroastrianism/Zoroastrianism1_Intro.pdf, p. 39.

2. See Philip G. Kreyenbroek, "Millennialism and Eschatology in the Zoroastrian Tradition," in *Imagining The End: Visions of Apocalypse from the Ancient Middle East to Modern America*, ed. Abbas Amanat and Magnus Bernhardsson (IB Tauris & Co Ltd, 2002), p. 35.

3. Again, see Philip G. Kreyenbroek, "Millennialism and Eschatology in the Zoroastrian Tradition," in *Imagining The End: Visions of Apocalypse from the Ancient Middle East to Modern America*, ed. Abbas Amanat and Magnus Bernhardsson (IB Tauris & Co Ltd, 2002), p. 35.

4. See "Sacrifice In Zoroastrianism," in *Encyclopædia Iranica*, http://www.iranicaonline.org/articles/sacrifice-i.

5. See Philip G. Kreyenbroek, "Millennialism and Eschatology in the Zoroastrian Tradition," in *Imagining The End: Visions of Apocalypse from the Ancient Middle East to Modern America*, ed. Abbas Amanat and Magnus Bernhardsson (IB Tauris & Co Ltd, 2002), p. 35.

6. See Prods Oktor Skjærvø, "Introduction to Zoroastrianism," http://www.fas.harvard.edu/~iranian/Zoroastrianism/Zoroastrianism1_Intro.pdf, p. 50.

7. Again, see Prods Oktor Skjærvø, "Introduction to Zoroastrianism," http://www.fas.harvard.edu/~iranian/Zoroastrianism/Zoroastrianism1_Intro.pdf, p. 50.

8. See Prods Oktor Skjærvø, "Introduction to Zoroastrianism," http://www.fas.harvard.edu/~iranian/Zoroastrianism/Zoroastrianism1_Intro.pdf, p. 22, 56.

9. See Philip G. Kreyenbroek, "Millennialism and Eschatology in the Zoroastrian Tradition," in *Imagining The End: Visions of Apocalypse from the Ancient Middle East to Modern America*, ed. Abbas Amanat and Magnus Bernhardsson (IB Tauris & Co Ltd, 2002), p. 38.

10. Again, see Philip G. Kreyenbroek, "Millennialism and Eschatology in the Zoroastrian Tradition," in *Imagining The End: Visions of Apocalypse from the Ancient Middle East to Modern America*, ed. Abbas Amanat and Magnus Bernhardsson (IB Tauris & Co Ltd, 2002), p. 38.

11. See Prods Oktor Skjærvø, "Introduction to Zoroastrianism," http://www.

fas.harvard.edu/~iranian/Zoroastrianism/Zoroastrianism1_Intro.pdf, p. 56.

12. See Philip G. Kreyenbroek, "Millennialism and Eschatology in the Zoroastrian Tradition," in *Imagining The End: Visions of Apocalypse from the Ancient Middle East to Modern America*, ed. Abbas Amanat and Magnus Bernhardsson (IB Tauris & Co Ltd, 2002), p. 39.

13. Again, see Philip G. Kreyenbroek, "Millennialism and Eschatology in the Zoroastrian Tradition," in *Imagining The End: Visions of Apocalypse from the Ancient Middle East to Modern America*, ed. Abbas Amanat and Magnus Bernhardsson (IB Tauris & Co Ltd, 2002), p. 39.

14. See Prods Oktor Skjærvø, "Introduction to Zoroastrianism," http://www.fas.harvard.edu/~iranian/Zoroastrianism/Zoroastrianism1_Intro.pdf, p. 57.

15. Again, see Prods Oktor Skjærvø, "Introduction to Zoroastrianism," http://www.fas.harvard.edu/~iranian/Zoroastrianism/Zoroastrianism1_Intro.pdf, p. 57.

16. See Philip G. Kreyenbroek, "Millennialism and Eschatology in the Zoroastrian Tradition," in *Imagining The End: Visions of Apocalypse from the Ancient Middle East to Modern America*, ed. Abbas Amanat and Magnus Bernhardsson (IB Tauris & Co Ltd, 2002), p. 36.

17. There is, incidentally, an alternative account according to which the wicked are eternally damned rather than purified.

18. See Prods Oktor Skjaervo, *The Spirit of Zoroastrianism* (Yale University Press, 2012), p. 171.

19. See Philip G. Kreyenbroek, "Millennialism and Eschatology in the Zoroastrian Tradition," in *Imagining The End: Visions of Apocalypse from the Ancient Middle East to Modern America*, ed. Abbas Amanat and Magnus Bernhardsson (IB Tauris & Co Ltd, 2002), p. 44.

20. This chapter relied heavily on Philip G. Kreyenbroek, "Millennialism and Eschatology in the Zoroastrian Tradition," in *Imagining The End: Visions of Apocalypse from the Ancient Middle East to Modern America*, ed. Abbas Amanat and Magnus Bernhardsson (IB Tauris & Co Ltd, 2002); Prods Oktor Skjaervo, *The Spirit of Zoroastrianism* (Yale University Press, 2012); Prods Oktor Skjærvø, "Introduction to Zoroastrianism," http://www.fas.harvard.edu/~iranian/Zoroastrianism/Zoroastrianism1_Intro.pdf; and Mahnaz Moazami, "Millennialism, Eschatology, and Messianic Figures in Iranian Tradition," http://www.mille.org/publications/winter2000/moazami.PDF. My intellectual labor was spent primarily compiling the extraordinary scholarship of these individuals into a single chronological narrative. Thanks to Philip Kreyenbroek for helpful comments on this appendix.

Appendix 3

Religion without God

·····

It's crucial to note that the influence of eschatology on world events goes far beyond the domain of explicitly *religious* belief. That is to say, history reports a number of "secular" eschatologies that have been widely accepted by the masses and, consequently, have had a nontrivial impact on the historical trajectory of civilization. While these may be "secular" in a strict sense, they share many important features with the eschatologies of religion, including most notably their epistemological basis in faith rather than evidence (see appendix 4). Consequently, I classify them as religious rather than secular, the latter term being reserved for proposed apocalyptic scenarios that derive from empirical observation and logical inference rather than mere ideological dogmatism.

One of the most influential eschatologies of this sort was developed by the nineteenth-century philosopher Karl Marx.* As Daniel Chirot and Clark McCauley observe in *Why Not Kill Them All?* "Marxist eschatology actually mimicked Christian doctrine."[1] According to the Marxist narrative, humanity began in a utopian condition, with no private property, classes, alienation, etc. But then private property and exploitation (sin) began to corrupt the world. Passing through different "dispensations"—the slave society, feudalism, capitalism—a messianic figure appeared (Marx) to show people a path back to utopia. In the end, "a final, terrible revolution will wipe out capitalism, alienation, exploitation, and inequality," and a special "elect" (the communists) will enter into the posthistory paradise of distributed wealth and abundance.[2] While Marx's critique of capitalism

* Perhaps as China emerges as an increasingly powerful world force in the coming decades, the end-times narrative of communism will become more relevant to discussions such as this. A future edition of this book might thus include an expanded section on the "secular" eschatology of Marxist ideology.

may still be worth reading (after all, capitalism *is* bound up with greed, exploitation, and alienation), his dialectical theory of history, with its attendant claims about the eschatological inevitability of communism, are demonstrably false. They're far closer in nature to the millenarian nonsense of dispensationalist Christians than the reasoned warnings of secular riskologists like Sir Martin Rees and Nick Bostrom.

Yet another "secular" eschatology with deeply religious motifs comes from Nazism. As Chirot and McCauley note in the same book (to quote them at length),

> It was not an accident that Hitler promised a Thousand Year Reich, a millennium of perfection, similar to the thousand-year reign of goodness promised in Revelation before the return of evil, the great battle between good and evil, and the final triumph of God over Satan. The entire imagery of his Nazi Party and regime was deeply mystical, suffused with religious, often Christian, liturgical symbolism, and it appealed to a higher law, to a mission decreed by fate and entrusted to the prophet Hitler.[3]

The point is that the millenarian impulse mentioned in chapter 1 extends far beyond the domain of religion *per se*. There are many instances in which "secular" systems of belief are ultimately based on faith no less than religious systems; some of these systems replace God with the State, a charismatic Dictator, or some other form of authority. Even Ray Kurzweil's Singularitarianism, with its belief in a "techno-rapture" that will usher in a millenarian transformation of the cosmos, leading to an unimaginable utopian state of ageless uploaded minds, exhibits strong religious parallels. To overcome the dangers of eschatology, we must be absolutely sure our beliefs about the future are properly proportioned to the evidence—and this leads us to the final section of this book, appendix 4.

Notes

1. Quoted in Steven Pinker, *The Better Angels of Our Nature: Why Violence Has Declined* (Penguin Books, 2011), p. 330.

2. Again, quoted in Steven Pinker, *The Better Angels of Our Nature: Why Violence Has Declined* (Penguin Books, 2011), p. 330.

3. Once more, quoted in Steven Pinker, *The Better Angels of Our Nature: Why Violence Has Declined* (Penguin Books, 2011), pp. 330–331.

Appendix 4

Thinking Clearly about the Big Picture

• • • • •

A Fundamental Difference

I mentioned in chapter 1 that the most crucial distinction between eschatology's two branches concerns not *what* these different narratives say about the future, but *why* they say it. In the case of religion, its stories in future tense are based almost entirely on faith in propositions acquired through revelations privately revealed to prophets with special access to the supernatural (often centuries or millennia ago). In the case of secular eschatology, its stories are based on evidence acquired through empirical observation of the natural world (plus truth-preserving and probabilistic inference). It's this difference that accounts for why secular eschatology ought to be taken seriously, while religious eschatology shouldn't. It accounts for why a secular riskologist shouting "Wolf!" ought to be heeded—perhaps with pupils dilated and heart racing—while a religious apocalypticist ought to be dismissed as a champion of nonsense.

In the present appendix, I want to flesh out these claims in more detail. There's probably no field of inquiry more fundamental to human flourishing than epistemology, since at the heart of this field is the paramount goal of becoming a *reasonable person*, and people who are reasonable are in a far better position to navigate a world littered with bad ideas than those who aren't. (And, to be sure, there are far more bad ideas out there than good ones.) As the atheist neuroscientist Sam Harris points out in an interview with the *Sun Magazine*, "No culture in human history ever suffered because its people became too reasonable or too desirous of having evidence in defense of their core beliefs."[1] This idea is just as applicable to individuals, societies, and even civilization as a whole.

But wanting to be reasonable is easier said than done. To understand what it means to be *able* to *reason*, one needs an elementary understanding

270

of Epistemology 101. The aim of this appendix is thus to offer a crash course in this field, providing some conceptual clarity to issues of truth, justification, belief, faith, and knowledge. If there's one thing philosophy can contribute to the world, it is, in my opinion, to give people the mental tools necessary to separate the good ideas from the bad ones. The following discussion frames everything else in this book, from our examination of the Simulation Hypothesis to the idea that a "clash of eschatologies" is driving some of the biggest conflicts of the age. The fact that this section comes at the end of the book (rather than being a chapter) has nothing to do with its importance—indeed, this is perhaps *the* most important part of the book, intellectually speaking. This section being an appendix is based purely on considerations of organizational coherence.

What Kinds of Things Are True?

Truth is not an esoteric phenomenon in the sense that the Standard Model of particle physics and the Hardy-Weinberg principle from evolutionary biology are. Truth makes an appearance in normal conversation all the time. Terms like "yes," "right," "correct," "absolutely," "agreed," and even "uh-huh" serve as affirmations of truth. Yet few of us have a solid grasp of what exactly truth *is*, of what truth *consists of*. That is, what exactly does it mean to *assert* that "Jesus will fight the Antichrist," "An asteroid will collide with Earth on May 5," and "Our universe began with a bang and will end in a whimper"? What does it mean to *agree* that any of these statements is correct?

The first step in understanding the nature of truth is to determine the sorts of things that can *be* true. Truth appears to be a property "had" by objects, just like *being funny* is a property "had" by Ricky Gervais, and *being tall* is a property "had" by giant sequoias. So, what kind of objects can "have" the property of being true? Let's begin simply: How about a chair? This may look like a silly suggestion, and perhaps it is, but philosophy has a way of surprising our untrained intuitions. In this case, however, the answer does appear to be "no." What would an argument between two people over whether a chair is "true" or "false" really amount to? It appears to be nonsensical.

What about, then, the *act* of drinking coffee? I have a habit of waking up every morning to a hot cup of joe—this much is correct—but could my action *itself* be called true? The answer again appears to be "no": an action can't be true or false any more than a chair. The philosopher Gilbert

Ryle famously referred to such misapplications of properties as "category errors." In the present context, applying truth to chairs or actions would be analogous to claiming that the concept of democracy weighs 20 pounds, or that time can literally crawl.

Philosophers have been working on the question of what things "have" truth for over 2 millennia, and the best answer they've come up with is this: the only objects in the whole universe capable of having the properties of truth or falsity are abstract entities called *propositions*. These are, as philosophers put it, the "bearers of truth." It's hard to explain what exactly a proposition is: like so many philosophical ideas, it appears graspable at first but turns out to be quite slippery. Philosophers typically characterize them as the *meanings* of sentences. To use a classic example from philosophy, "Schnee ist weiss" in German and "La neige est blanche" in French both express the same proposition as the English sentence "Snow is white." By convention, we'll write propositions in italics from now on, as in, *Snow is white*.

As the meanings of sentences, propositions *represent* some configuration of reality, some state of affairs in the world. *The Islamic State wishes to start a war in Syria*, for example, represents one possible configuration of reality, and we can express this proposition using the English alphabet as follows: "The Islamic State wishes to start a war in Syria." *Jesus is God* represents another possible configuration of the world, and speakers of Spanish can express it as "Jesús es Dios." *Philosophy is hard* represents yet another configuration, and speakers of Portuguese can express it as "Filosofia é difícil." Propositions are why sentences in different languages can have the same meaning.

This is a fairly technical account of propositions. There's a much easier way to understand them, though. Borrowing from cartography, we can say that propositions are nothing more than *maps of reality* (or some region therein). But unlike literal maps made of paper or displayed on a computer screen, propositional maps exist in the mentalistic space between our ears. Thus, *World War II ended in 1945* is a map that I have inside my head according to which World War II ended in 1945. The proposition *Berlin is in northern Africa* is a map that I *could* have in my head, but don't—and you shouldn't either, because Berlin is in Germany. There are an infinite number of such maps floating above your head, any one of which you could reach up and grab ahold of. Rephrasing a claim from above, the particular sentences "Schnee ist weiss," "La neige est blanche," and "Snow is white" all mean the same thing because each refers to the same map of reality.

Q: What are the only things in the universe that can be true or false?

A: Propositions.

Q: What are propositions?

A: The meaning of sentences. You can think of them as maps in the mind that represent some aspect of reality.

But What Makes It True?

If truth is a property of mental maps in our heads—maps that represent different configurations of the world—then what determines whether a map is true or false? Why is *The earth orbits the sun* true, whereas *The sun orbits the earth* is false? Thinking about propositions in cartographic terms makes the answer to these questions easy to understand. Consider a "political map" of the world, which outlines the borders of countries. Under what conditions would it be considered accurate? The answer is, of course, that the map is accurate when what it *says* the world is like lines up sufficiently well with what the world is *actually* like. An accurate—or we could say "true"—map of the world's countries should represent Russia being north of China, China being east of India, India being northeast of South Africa, and South Africa being south of France. The crucial point is that there must, for the map to count as accurate, exist some kind of *correspondence* between (a) the map (a representation of reality) and (b) reality itself.

With respect to propositions, then, *The earth orbits the sun* counts as true when it accurately corresponds to the way our solar system is actually configured. We might say that it's true when it can be *mapped onto* reality in a one-to-one manner. The proposition *Berlin is in northern Africa*, in contrast, cannot be mapped onto reality with any degree of accuracy, and this is why it's false. Truth thus consists in a correspondence between mental maps and the worldly regions they purport to represent. This is called the *correspondence theory of truth*, and according to a 2013 survey by David Chalmers and David Bourget it's the most popular theory of truth among philosophers today.[2]

It's important to notice that, on this theory, truth has nothing to do with what we, or any other sentient creatures in the cosmos, *believe* about the world. That is to say, maps are true or false *on their own*, independent of whether anyone accepts, rejects, or has ever even considered them as a possibility. The map *There is a 1-km-wide asteroid heading toward Earth* is true or false independent of what any human currently believes about asteroids. As philosophers put it, propositions are true or false in a "mind-independent" way: their status as *being true* or *being false* has nothing to do with what goes on in anyone's mind. The sun, for example, is 93 million miles from the earth whether anyone believes this or not. Truth is an *objective* phenomenon.

Q: What is truth?

A: A relation of correspondence between maps of reality and reality itself. It has nothing to do with what we think of the maps, and in this sense truth is objective, or "mind-dependent."

How Do Beliefs Fit into This Picture?

So far we've mentioned beliefs several times. But what exactly does it mean to believe something? According to many philosophers, belief consists of a *relation* between exactly two rather different things. To illustrate this relation, let's dissect a simple sentence expressing a Christian belief, namely "I believe that Jesus healed a blind man with mud made from spittle" (Mark 8:22-25). First we have the subject "I," followed by the relational term "believe." We can already see, then, that the belief relation is between a *speaker* (or, more generally, a *cognitive agent*) and something else. Dissecting the sentence further, we find "believe" followed by the word "that." This word is important because it serves to introduce a particular description of one possible configuration of reality—i.e., it introduces a *proposition*. Thus, on the other side of "believe" is a sentential structure that refers to a mental map: *Jesus healed a blind man with mud made from spittle*. So, a belief is an abstract relation between people and propositions.

This is why philosophers classify beliefs as a kind of "propositional attitude": it constitutes a particular *stance* one can take toward a proposition, namely the stance of "cognitive assent." In other words, a belief is what happens when a cognitive agent *accepts* a given map as being an accurate representation of reality. It's what happens when one comes to the conclusion that a proposition has the property of *being true*. In contrast, a desire is also a propositional attitude, but it doesn't involve accepting a proposition as true; rather, it involves *wanting* a proposition to be true, whether or not it is. Many philosophers would argue that much of human behavior (if not all) can be explained in terms of the interaction between beliefs and desires. For example, the desire to eat pizza plus the belief that there's a pizza in the freezer leads to the behavior of turning on the oven and opening the freezer. Given the intuitiveness of such explanations (most of us naturally think about others' behaviors in terms of beliefs and desires), this approach is typically labeled "folk psychology."

Q: How do beliefs fit into this?

A: A belief is what happens when a mind accepts that a proposition is true. In the absence of acceptance, one has no belief.

When Should One Believe something?

What are the conditions under which psychologically accepting a proposition as true is justified, warranted, rational, or reasonable? (Note that these words are all more or less synonymous in this context.) When should one accept a mental map as true? The best answer that philosophers have come up with is that a cognitive agent-proposition relation is justified when, and only when, it's *proportioned to the evidence*. To quote the twentieth-century philosophers WV Quine and SJ Ullian, "Insofar as we are rational in our beliefs, the intensity of belief will tend to correspond to the firmness of the available evidence."[3] Thus, the more compelling the evidence, the stronger the belief should be held; the less compelling the evidence, the weaker it should be held. This theory of justification is called *evidentialism*.[4]

One way of framing the idea is this: epistemology, in a manner of speaking, demands that one must always have *reasons* for believing; evidentialism then claims that having evidence is what it *means* to have reasons. Like folk psychology, evidentialism captures our "pre-theoretic" intuitions quite nicely: most people behave as though they're evidentialists in almost every domain of life—religion being the most conspicuous and important exception, as we'll explore more below. For example, if you want to find out who won the hockey game last night, you don't close your eyes and try to *intuit* the answer, nor do you (presumably) *pray* for the final score to be revealed to you by an angel of God. Rather, you're likely to turn on the TV or check the Internet for the answer, or perhaps ask a trusted friend. This is nothing more than evidence gathering. It's based on the idea that "if I want my belief about last night's hockey game to accurately correspond to reality, I need the belief relation to be justified by some evidence."

Evidentialism is, therefore, not merely a "normative" theory about how one *ought to* justify one's worldview but a propitious "descriptive" account of how people *often do* support their assertions. (While many theories in philosophy and science offend our untrained intuitions—how unintuitive is relativity theory, quantum mechanics, and Darwinian evolution!—this is an instance of our intuitions and the best proposed theory aligning without much tension.) It's also at the core of secular eschatology. What makes the risks posed by nuclear conflagrations, advanced biotechnology, and the creation of a superintelligence, for example, worth taking seriously is the fact that they're all possible scenarios inferred from the evidence. For this reason, *just as religious eschatology is an appendage of theology, secular eschatology is a field of science*: its gloom-and-doom stories are nothing more than hypotheses about the future derived through the empirical method of science.

Some Qualifications

Although evidentialism is fairly straightforward, there are a number of qualifications that need to be made. One is that justification stems not from having *some* evidence for a belief but from the *totality of evidence available* at the time. There are, for example, many real-world situations in which people are confronted by two incompatible maps of reality that they must decide between. Let's say that map A contradicts map B, and A is supported by three "pieces" of evidence (ignoring the oddness of "pieces"). Does this make A reasonable to accept? The answer depends entirely on how much evidence B has in its favor. If it turns out that B is supported by twenty "pieces" of evidence, then it would be *unreasonable* to accept A, even though it has some *positive* evidential support.[5] The key idea is that one must scan the whole pool of evidence and choose a theory based on its *relative evidential support*. This is, incidentally, where a lot of conspiracy theories go wrong: they tend to focus on a few pieces of evidence that seem to support a theory, without looking at the situation as a whole. Perhaps there is some evidence compatible with 9/11 being an inside job perpetrated by the US government, but examining the entire situation, one finds that the totality of evidence overwhelmingly supports the opposite conclusion. 9/11 was *not* an inside job, but a terrorist attack perpetrated by al-Qaeda.*

The second qualification is that evidence cannot *guarantee* the truth of any belief, *ever*. The only instances in which absolute confidence is warranted are those involving (what we'll broadly term) *logical truths*. A circle cannot have corners, for example, and we can know this with the highest degree of certainty possible. Simply grasping the concept of a circle is enough to apprehend this knowledge. Similarly, a Penrose triangle could never actually be built, a bachelor, as dictionaries currently define the word, could not possibly have a wife, and a God who's bearded and clean-shaven at the exact same moment could never exist, given the Law of Noncontradiction. These are truths that we don't need to explore the world to affirm: they're justified not for empirical, but logical reasons. They are, as it were, *self-evident*, meaning that the evidence comes from merely *understanding* the relevant concepts.

But, and this is a crucial point, *whenever a proposition can't be justified by logic alone, our only option is to examine the evidence.* The snag is that the totality of our best current evidence is not the same as the totality of all the *possible* evidence: in most domains, not every piece of evidence has been uncovered; research is ongoing. This is precisely why science produces inconsistent results over time. For example, space and time were once thought to be completely separate entities, but physicists since Einstein have believed that they form a continuum. Similarly, a blood-lead level of less than 10 micrograms per deciliter of blood in children was once thought to be safe, but toxicologists today believe that no amount of lead in the body is safe. And in 1960 only about one-third of doctors in the US believed that smoking causes cancer, while virtually everyone today—doctors and patients alike—agrees it does.[6]

The reason for such flip-flopping is quite simply that the relevant pool of evidence grew as a result of further scientific research: experimentation in the "hard" sciences and investigation in the "soft" sciences. This is the

* Notice also that there's a difference between evidence being compatible with a hypothesis and evidence **supporting** it. The observed evidence is compatible with the existence of invisible, perfectly silent unicorns all around us, but our observations of absent unicorns certainly doesn't give us reason for thinking they're real.

epistemologically *good* kind of flip-flopping, the kind of flip-flopping we *want* our politicians to engage in. It's flip-flopping motivated by evidential considerations, rather than political, social, economic, or religious factors.

By analogy, devising a hypothesis on the cutting edge of science is like guessing the picture printed on a 1,000-piece jigsaw puzzle *with only a few pieces to look at.* These pieces might show something that looks like the green leaves of a tree, and this highly limited amount of evidence—the best available at the time—might lead you to propose that the image is a forest. Putting more pieces together, you realize that it's not of the forest but the inside a house (you see walls and a rug), and consequently you put forth a new hypothesis: the green leaves are a house plant in the corner of a room. Later on, you assemble 950 pieces and, with this nearly complete image, you realize that the picture is actually of a living room with a *painting* of the forest hanging on the wall (and no plant in the corner). In this example, then, you made a series of evidence-based guesses, each of which was justified and reasonable when you made them, even though two of them later turned out to be false.

It follows that as science accumulates larger amounts of evidence, it's always possible for some critical, contradictory datum to turn up. There's nothing requiring that all future evidence is consistent with all past evidence. Evidence is, rather, a *guide* to the truth—nothing else. It sometimes points us in the wrong direction, but over time it homes in on the truth like a heat-seeking missile. This is why, as philosophers have noted, insofar as people are rational, their beliefs about reality *will tend to converge over time.*˙ Science is a perfect example of this convergence, given that the entire enterprise is founded upon the gathering and analysis of belief-supporting evidence. From one side of the planet to the other, one finds an extraordinary amount of *agreement* about the most fundamental scientific issues, including the laws of physics, the origin of the cosmos, the table of elements, the mechanisms of evolution, the anatomy of horses, the age of the earth, the structure of the human brain, the timing of the

---

* As the Princeton philosopher Thomas Kelly writes, "To the extent that individuals and institutions are objective in this sense, we should expect their views to increasingly converge over time: as shared evidence accumulates, consensus tends to emerge with respect to formerly disputed questions. Objective inquiry is evidence-driven inquiry, which makes for intersubjective agreement among inquirers." See Thomas Kelly, "Evidence," in **The Stanford Encyclopedia of Philosophy**, ed. Edward Zalta (Fall 2014 edition), http://plato.stanford.edu/archives/fall2014/entries/evidence/.

dinosaurs' extinction, the distance of Earth from the sun, *and so on*. From China to Chile, Brazil to Belgium, the US to the UAE, scientists are in almost complete agreement about phenomena like these.

It's worth noting just how much this contrasts with the situation of religion, which is not evidence-based. In this domain, one finds an extraordinary amount of *disagreement* about the most fundamental theological issues, including the nature of God, the fate of the human soul, the legitimacy of prophets, the importance of different rituals, the ultimate fate of the universe, *and so on*. The reason is simply that for science, beliefs are the *destinations* of a journey whose course is determined entirely by the best available evidence; for religion, beliefs are the *points of departure*. Whereas science aims to follow its evidential guide wherever it leads (even when this entails discarding deeply held prior beliefs), religion expends its energy trying to defend the "truths" that its traditions started with— "truths" privately revealed to prophets by the supernatural. In a phrase, science is based on *fallibilism*, or the possibility of always being wrong, whereas religion is based on *dogmatism*, or the certainty that one's initial beliefs are eternally correct. This is, from the perspective of epistemology, the fundamental difference between science and religion—and therefore between secular and religious eschatology.

A final qualification of evidentialism has to do with the sorts of phenomena that can count as evidence. Consider a fascinating story: in 2008, the neurosurgeon Eben Alexander spent 7 days in a coma after contracting bacterial meningitis. While on the precipice of death, his soul embarked on an ecstatic adventure through the mystical realm of the afterlife, flying on the wings of a butterfly over kaleidoscopic panoplies full of flowers and trees covered in blossoms. Eventually, Dr. Alexander met God himself, as detailed in his best-selling *Proof of Heaven: A Neurosurgeon's Journey into the Afterlife*. What's interesting is that Alexander sees his experiences as providing not just evidence for *him*, but evidence for *everyone else* too. That is to say, he takes his experiences to have yielded the sort of evidence that can justify cognitive assent to propositional maps like "Heaven exists" and "There is a life after this one." As a result, *you* are unreasonable if, in the face of *his* spiritual adventure, *you* continue to believe that God isn't real and our souls aren't immortal.

Does Dr. Alexander's experience constitute genuine evidence? No, probably not for him and definitely not for us. The reason is that *good evidence is checkable*. Checkability matters because it's our best way to guard against confounding factors that often arise from the fragility and fallibility of our minds. It is, indeed, far too easy for humans to be mistaken, self-deceived, deluded, or tricked into thinking something's the case when it's not. Checkability protects us from becoming a victim of these epistemic threats. In some cases, of course, we may never be able to check the claim *ourselves*, but the claim is at least checkable *in principle*. We then rely on trustworthy sources to convey the truth to us and can even *check* these sources themselves to ensure that they're reliable. In other cases, though, there's no way to check *in principle*. This is precisely the case with the subjective "feelings" had, "voices" heard, and "presences" felt by religious people all over the world, from North America to the Middle East. It also applies to many instances of revelation reported by the supposed prophets of God, including Adam, Noah, Moses, Abraham, Ishmael, Solomon, Ezekiel, Jesus, the Apostle Paul, the Apostle John, Muhammad, Joseph Smith, David Koresh, Shoko Asahara, Vissarion, and Brian David Mitchell (the man who kidnapped Elizabeth Smart because, apparently, God told him to do it).

One of the great epistemological breakthroughs of modern science is its demand that every theory must be *entirely justified by checkable evidence*. Thus, for a theory to be fully accepted by the scientific community, it must be built upon evidence that different people with different perspectives in different places at different times can independently verify. The benefit of this brilliant strategy is that no one is forced to take anyone else's word for it: the option of *seeing for yourself* is, in principle, always open, even if in practice it's sometimes impossible due to educational, intellectual, and technological constraints.

Another problem with "evidence" acquired through subjective experience is that it's so *consistently contradictory* as to cast serious doubt upon such experiences as a reliable means for gathering belief-supporting evidence. The feelings of divinity had by Muslims, for example, are no less phenomenologically real than those had by Christians, yet they support a worldview that's mutually incompatible with Christianity. And the revelations had by Paul were probably no less psychologically real than those had by Muhammad, yet they fit into a religious tradition that's fundamentally inconsistent with Islam. Think for a moment about what this means: it implies that *subjective experiences of God, including supernatural revelations, are highly unreliable means for acquiring truths about reality*.

From a statistical point of view, this conclusion follows *even if* one of the world's religions is true. Imagine for a moment that the Christian worldview, accepted by about 31% of the global population, constitutes an inerrant and complete account of the deepest nature of reality. Since Christianity is incompatible with Islam, and since 23% of the world is Muslim, this means that 23% of the global population lives their lives, everyday, feeling an intimate connection with a God no less imaginary than a child's invisible playmate. The situation is even worse than this, though, since it entails that another 15% of humanity (Hindus), plus 7% (Buddhists), plus 6% (folk religionists), plus 0.2% (Jews), and so on are all similarly fixated on completely false "truths." It follows that the *mathematical majority* of religious experiences reported by sincere believers around the world don't actually correspond to any external reality. In concrete terms, if one had to place a bet on whether a given experience reported by a given believer was veridical or not, you'd be much more likely to win if you put your money on it being a delusion.[7]

An exactly similar thing could be said about revelation: *even if* one of the world religions were true, it would still be the case that the large majority of revelations had by prophets throughout history were actually delusions. Statistically speaking, revelation as a method for acquiring truths about the universe has an *absolutely terrible track record*: it has almost always (if not always) led to falsities rather than truths. This is why religious exclusivism (the view that one, and only one, religion is correct) is a *self-defeating* position: believing that, for example, Christianity is true itself gives one reason for thinking that it's false. Why? Because religious exclusivism entails that the majority of religious experiences and reported revelations had by equally sincere believers throughout history have not been veridical, and therefore that these modes of accessing truth are, at best, statistically unreliable; yet the foundation of Christianity is revelation and religious experience. Ergo, the "truths" of Christianity are probably false. The exact same argument applies to Islam, Mormonism, the Baha'i Faith, and every other system based on special revelation. Believing that a revealed religion offers the one and only truth gives one reason for doubting its veracity.

The method of revelation contrasts strongly with that of observation, which is favored by science. Whereas revelation has proven itself to yield far more falsities than truths—if any at all—observation has proven itself to be a highly dependable means for acquiring exactly the sort of evidence needed to justify beliefs. Although observation is by no means perfect—just stick a pencil in a cup of water and watch it bend in half!—we can, as the philosopher of science Peter Godfrey-Smith points out, use the empirical method of science to look at our situation "sideways-on" to "learn *what kind of reliability we have* in our attempts to know about the world," an approach that "can be applied to inferences and modeling strategies in science itself."[8] In other words, we can construct theories that *take into account* the ways in which observation is deceptive, thereby making our theories that much more robust.[9]

Even putting this insight aside, the fact is that observation doesn't yield anything close to the sort of "consistent contradiction" that's become a hallmark of revelation. With a little training, virtually *anyone* with two eyes can examine a mandible from a particular layer of the earth and identify it as belonging to an australopithecine, or distinguish between the endoplasmic reticulum and the Golgi apparatus through a microscope. Observation is perhaps the most central aspect of the scientific method, and in this way the heart of science is a philosophical position called *empiricism*. Whereas evidentialism answers the question "*Why* do you believe X?" ("Because the totality of evidence favors X"), empiricism answers the question "*How* do you know X?" ("I made an observation and acquired evidence that favors X").

To summarize the many important points of this section, a justified belief is one based not just on the totality of evidence available at the time but on evidence that can be publicly checked, at least in principle, by other parties. Checkability enables one to be confident that a piece of "evidence" wasn't acquired by someone who made a mistake, succumbed to self-deception, sank into delusion, or was tricked by another in a hoax. Religion foolishly rejects this safeguard: entire systems of violently incompatible convictions are founded on prophets saying—assuring you, with only force or charisma to back themselves up—that *they* have some special access to divinely revealed truth. Since you can't check that any of these individuals are right, your only option is to adopt the stance of *faith* with respect to their claims. And this brings us to the final section of this chapter.

Q: *How can you know that a proposition is true?*

A: *When it comes to maps of reality, you can't ever know with absolute certainty. The exception comes from logical claims like*

"Are there circles with corners?" and "Could there exist a God who's simultaneously both bearded and clean-shaven?"

Q: When should you believe a proposition?

A: When logic can't provide the answer, our only option is to assign a probability of truth based on the evidence. Specifically, the probability should be based on the totality of checkable evidence available at a given time. More evidence equals more confidence in the proposition's truth; less evidence equals less confidence.

Knowledge and Faith

What does it mean to *know* something? If a religious person holding a sign that reads "The end is near" tells you that next Wednesday a cosmic catastrophe is going to destroy the earth, and if next Wednesday this actually happens, would you say that she *knew* this was going to happen? What about situations in which you have *a lot* of evidence for a belief but that belief turns out to be false? As mentioned above, there was a time when our best evidence suggested that space and time are distinct, but we now know that this is false. Or imagine our ancient ancestors from the Paleolithic, observing the sun traverse the sky each day, while the earth beneath them feels stationary. This is all the evidence they had to go by at the time, and it strongly suggests (wouldn't you agree?) that the sun revolves around the earth. Given that their *geocentric* ("Earth-centric") beliefs were based on the best available evidence, could these beliefs have counted as knowledge?

To answer these questions, we need a theory of knowledge. The best one to date was, amazingly, given to us by the ancient Greek philosopher Plato. Although there appear to be some problems with it—thanks to the contemporary philosopher Edmund Gettier, who overturned more than two millennia of thinking with a single, 3-page paper published in 1963—no better alternatives have been proposed. (Calling all young philosophers!) The Platonic theory states that knowledge consists of three *necessary and sufficient* conditions. This means that if even *one* condition is lacking, you ain't got knowledge; and just as soon as all three are satisfied, you can't help but have knowledge. Necessary and sufficient conditions are ways of specifying a phenomenon's *essence*. With respect to big-picture hazards, for example, *being global* is the sole necessary and sufficient condition. A risk thus falls into this category "if and only if" it possesses this property.

The first condition that Plato identifies is that to know X, one must also *believe* X. It doesn't make sense to talk about knowing something that you don't believe. The sentence "I know the earth is round but I don't believe it" is, quite clearly, confused. (So is the sentence "I know evolution is true but I don't believe it.") Knowing entails believing—although, as we'll see, believing doesn't entail knowing. The second condition is that X must be epistemologically justified. This is why the religious person's prophecy (above) doesn't count as knowledge, even though it was accurate. The third condition is that X must be true. No matter how much evidence there might have been for the earth being stationary at some point in the past, it couldn't have counted as knowledge because it's false. This is why our Paleolithic ancestors' belief in geocentrism didn't count as knowledge, even though it was (or may have been) evidentially justified. On the Platonic model, knowledge is a tripartite phenomenon consisting of a psychological (belief), an epistemological (justification), and a metaphysical (truth) component.

Faith is similar to knowledge in certain respects, and diametrically opposed to it in other respects (see table D-A). We should note right away that there are two distinct, common uses of the word "faith," only one of which concerns us here. On the one hand, Christians talk about having faith *in* God, Muslims having faith *in* Allah, Zoroastrians having faith *in* Ahura Mazdâ, and so on. This sense of the word is roughly synonymous with *trust*, and it's often introduced by the preposition "in." Philosophers refer to it as the *fiducial* meaning of the term, since the Latin word for trust is *fiduci*.[10] On the other hand, faith is used to affirm the truth of a proposition. Thus, Christians have faith *that* God exists, Muslims have faith *that* Allah exists, Zoroastrians have faith *that* Ahura Mazdâ exists, and so on. This sense of faith turns it into a species of *belief*, often preceded by the word "that" (exactly as in the case of propositional attitudes, as mentioned above). We can call this the *epistemological* meaning of the term—it's the much more philosophically interesting and important use of faith by religious believers, and it appears to be primary to all other meanings, since trusting *in* X doesn't make sense if one doesn't believe that X is real in the first place.

Thus, both knowledge and faith are kinds of belief. This is their common denominator. But there are two crucial differences between them: first, unlike knowledge, faith doesn't *require* that the accepted proposition be true. We know that Zeus is a fiction of Greek mythology, but this doesn't mean that one can't still have faith that he exists. It's entirely possible to have faith in known falsities. On the flip side, a proposition accepted out of faith doesn't necessitate that it's false. If it turned out that Zeus *is* real, this wouldn't affect the status of the ancient Greeks' faith in his existence. In a phrase, whether or not an instance of belief counts as faith is independent of whether or not the proposition believed is true or false. It could be either.

The second difference is even more significant. Whereas a belief must be justified to count as knowledge, the exact *opposite* is true of faith. This is, in fact, the defining feature of faith: a belief must *lack* supporting evidence for it to count as faith-based. Indeed, if evidence could be adduced to justify religious propositions like "God is three persons in one," "Jesus was born of a virgin," "The Antichrist will invade Israel and destroy the temple," "The Twelfth Imam is alive right now but in hiding," "Jesus will descend over east Damascus with two angels by his side," and "Hell will be a place of eternal torture," then the believer wouldn't need faith to accept them, and religion would be science. Faith is what the faithful use to fill the box of "reasons for believing" when no good, checkable evidence can be found. When evidence *can* fill the box, faith becomes superfluous.

One might think that this definition is tendentious—a semantic subterfuge of the atheist to back the religionist into a conceptual corner. But it's not just secular philosophers who accept it. The epistemological sense of faith as belief without evidence is the standard point of departure for discussions in the philosophy of religion, the epistemology of religion, certain branches of theology, and other related fields.[11] It's this definition that scholars who study the nature of religious belief, independent of their personal persuasions, take to be the starting point of discussion. This is why philosophers of religion and theologians have spent so much time and mental energy attempting to devise accounts of faith that aren't *obviously* unreasonable—that don't immediately fail by opening up the floodgates to all sorts of crazy beliefs.

Table D-A. Knowledge vs. Faith

	KNOWLEDGE	FAITH
Belief	Yes	Yes
True	Yes	Yes or No
Justified	Yes	No

Accepting a belief out of faith is epistemological foolishness. This is why the secular branch of eschatology has a completely different standing than the religious one. Yet far more people around the world are focused on the narratives of religion than the doomsday scenarios put forth by evidence-minded scientists. The ultimate aim of rational beings like ourselves is to get the interconnected networks of beliefs in our heads— our worldview—to maximally align with the reality outside our heads. Following the evidence, with logic and probability in our toolbox, is the very best way to do this.

Notes

1.   See Bethany Saltman, "The Temple of Reason," *Sun Magazine*, September 2006, http://thesunmagazine.org/issues/369/the_temple_of_reason?page=1. Harris has mentioned this idea in a large number of publications.

2.   See David Bourget and David Chalmers, "What Do Philosophers Believe?" *Philosophical Studies* 170 (2014): p. 16.

3.   See W. V. Quine and J. S. Ullian, *The Web of Belief* (McGraw-Hill Humanities/Social Sciences/Languages, 1978).

4.   Reliabilism asserts that a belief is justified when and only when it was produced by a "reliable process." Some Christian philosophers have taken this theory and posited—following John Calvin—a special cognitive module, called the *sensus divinitatis*, that's capable of "sensing the divine." Because this module is reliable, they argue, the beliefs it produces about God are therefore justified. Unfortunately, as the Christian philosopher Alvin Plantinga argues, this module doesn't work properly in some people as a result of sin. Many epistemologists find the notion of a *sensus divinitatis* extremely unsatisfying.

5.   Incidentally, Occam's razor—the heuristic according to which one should always prefer simplicity over complexity—comes into play when a situation is such that A and B are backed by exactly the same evidence, or evidence with the same strength. In this case, you can't use evidence to arbitrate between competing theories, so you have to turn to something else: simplicity. The reason is that there are less ways for simpler propositions to be wrong than complex ones. As Graham Oddie puts it in a *Stanford Encyclopedia of Philosophy* entry on "Truthlikeness," the "degree of informative content varies inversely with probability—the greater the content the less likely a theory is to be true."

6. See Robert Proctor, "The History of the Discovery of the Cigarette-lung Cancer Link: Evidentiary Traditions, Corporate Denial, Global Toll," *Tobacco Control* 22, no. 62 (2012).

7. This is related to, but distinct from, John Loftus' influential "Outsider Test for Faith," according to which (to quote Loftus) "the best way to test one's adopted religious faith is from the perspective of an outsider with the same level of skepticism used to evaluate other religious faiths." The reason is that (a) "rational people in distinct geographical locations around the globe overwhelmingly adopt and defend a wide diversity of religious faiths due to their upbringing and cultural heritage," (b) "it seems very likely that adopting one's religious faith is not merely a matter of independent rational judgment but is causally dependent on cultural conditions," and (c) "hence the odds are highly likely that any given adopted religious faith is false." See John Loftus, "The Outsider Test for Faith *Revisited*," in *The Christian Delusion: Why Faith Fails*, ed. John Loftus (Prometheus Books, 2010), p. 82.

8. See Peter Godfrey-Smith, *Theory and Reality* (University of Chicago Press, 2003), p. 223.

9. Note that, using science, we can also take into account the many cognitive biases mentioned in the previous chapter. By studying how our minds can "reason" in faulty ways, we can better defend against such tendencies.

10. See John Bishop, "Faith," in *Stanford Encyclopedia of Philosophy*, ed. Edward Zalta (Fall 2010 edition), http://plato.stanford.edu/archives/fall2010/entries/faith/.

11. See, for example, John Bishop, "Faith," in *Stanford Encyclopedia of Philosophy*, ed. Edward Zalta (Fall 2010 edition), http://plato.stanford.edu/archives/fall2010/entries/faith/; Charles Taliaferro, "Philosophy of Religion," in *Stanford Encyclopedia of Philosophy*, ed. Edward Zalta (Winter 2014 edition), http://plato.stanford.edu/archives/win2014/entries/philosophy-religion/; and Peter Forrest, "The Epistemology of Religion," in *Stanford Encyclopedia of Philosophy*, ed. Edward Zalta (Spring 2014 edition), http://plato.stanford.edu/archives/spr2014/entries/religion-epistemology/.

About the Author

· · · · ·

Phil Torres is an author, an Affiliate Scholar at the Institute for Ethics and Emerging Technologies, and founder of the X-Risks Institute for the Study of Extremism (www.risksandreligion.org). His background is in neuroscience and philosophy, and he has published in academic journals such as *Erkenntnis*, *Metaphilosophy*, *Foresight*, the *Journal of Future Studies*, and the *Journal of Evolution and Technology*. He's also a musician whose songs have been featured in commercials.